CATCH A ROCKET PLANE
More Tales From the Cutting Edge, and Beyond

DR. ROBERT F. BRODSKY

TILE BY ROBERT D. BRODSKY,
RABBIT ARTWORKS, SANTA FE, NM,
CIRCA 1978

CATCH A ROCKET PLANE
MORE TALES FROM THE CUTTING EDGE, AND BEYOND

For information contact the author at 650 W. Harrison Ave. Claremont, CA 91711; rfoxbro@aol.com; or (909) 626 2050

Copyright © 2012 Dr. Robert F. Brodsky
All rights reserved. No part of this publication may be reproduced, stored in a retrieval system, or transmitted, in any form by any means, electronic, mechanical, photocopying, recording, or otherwise without written permission of the author

ISBN: 1467972908
ISBN-13: 9781467972901
Library of Congress Control Number: 2011961705

Printed in the United States of America

DEDICATION

TO MY WIFE AND FAMILY AND MANY FRIENDS THROUGHOUT THE WORLD WHO HAVE SEEN ME SAFELY THROUGH A WONDERFUL LIFE'S JOURNEY; AND TO MY OLD FRIENDS AND COMPATRIOTS – AND THEIR SPOUSES – WHO JOINED IN THE EXCITEMENT OF DESIGNING AND TESTING OUR EARLY NUCLEAR BOMBS AT SANDIA CORPORATION. MOST, I FEAR, WITH THE EXCEPTION OF MARY JO AND PEP, ARE GONE – BUT NONE ARE FORGOTTEN, AND WILL NOW BE NAMED:

ALAN POPE	WARREN CURRY	ED CLARK
GEORGE HANSCHE	KEN ERICKSON	PETE PETERSEN
PAUL ROWE	HAL VAUGHN	MARY JO VAUGHN
BOB BLACK	CONRAD NELSON	BOB PEARCE
'MAC' McAVOY	BOB HENDERSON	BOB POOLE
CARRIE PUMPHREY	EATON DRAPER	BILL GARDNER
BILL PEPPER	NATALIE BRADLEY	JIM SHARP
'HAP' HAZARD	MEL MERRITT	JIM SHREVE
KEN CROWDER	OSSIE RITLAND	RANDY MAYDEW
EMILY EDWARDS	'MICK' MICHAELIS	VINEY STRANCE
BILL PUMPHREY	JIM STARK	HARRY EVANS

Other Books by Robert F. Brodsky

On the Cutting Edge (Gordian Knot Books, 2006)
Songs My Mother Never Sang to Me (FOXBRO Press, 2008)
A Pilgrim Muddles Through (FOXBRO Press, 2009)
The World in a Jug (FOXBRO Press, 2010)

CONTENTS

Introduction/Acknowledgement vii
Preface   The Engineering Life ix
Chapter 1   The Making of an Engineer 1
    Learning to fly at Cornell; Navy adventures during the War; Hot Jazz and grad school
Chapter 2   Atom Bomb Stories 27
    The dawn of the nuclear age in New Mexico; Settling in at Sandia; The people who made the Bomb
Chapter 3   The Space Age Cometh 71
    Getting to California and then into Space; Living it up in Paris; Pioneering space adventures
Chapter 4   La Vie Academe 105
    Ames, Iowa in the '70s; Bringing Space to Academia; The Faculty Improvement Leave sabbatical in the South Bay
Chapter 5   Space by the Sea 119
    Working at TRW- Teaching at USC; Fighting for the Space Lifeboat, Visiting Professoring in Haifa
Chapter 6   Snapshots from the Turn of the Century 147
    A survey of lecture reviews made by the movers and shakers in The period 1997 to 2009
Chapter 7   Winding down at Work 209
    Assessing aerospace engineering academic programs; discussing the fine art of teaching at the University level; retirement stories
Chapter 8   Ain't Retirement Grand 247
    The 'expert witnessing' game; Starring on stage, screen and radio; 'round the World, the 70th Reunion
Index 331
Picture Listing 334

(NOTE: THE INDEX (P. 331) AT THE END OF THE BOOK GIVES DETAILS OF THE CONTENT OF EACH CHAPTER; A LISTING OF ILLUSTRATIONS FOLLOWS IT (P. 334))

INTRODUCTION / ACKNOWLEDGEMENT

"*Rocket Plane*" – my fifth – may be my last book. I've run out of stories that I want to tell and which would interest the reader and my extended family. I achieved 86 in May, 2011, and my wife wants us to move to a retirement community. It's not that I am weary or burned out, or that my memory has failed - it's just that the only message I have left is my disappointment with our political system and how it has let human greed ruin our immediate part of the universe. Who wants to hear about that! And, no one seems interested in seriously pursuing what I know to be the earth's inevitable future – going on to the Hydrogen Economy.

This book has turned out to be a memoir that fills in some of the spaces in my both professional and non-professional life that were not covered in my first book, "*On the Cutting Edge*". It provides background material covering a broad panorama of a time period (1944-2010) dominated by technical advances made in the worlds of Aerospace, Electronics and Education, in which I was a participant. The book title, as you will see in the Footnote* below, comes from the words of a rare 78 rpm record in my collection: "Chicago Woman Blues". The picture on the 'inside' cover comes from a tile that our ceramist son, Robert D. – proprietor of the *Rab-*

Ruby Smith, "CHICAGO WOMAN BLUES", parts 1 & 2, Harmonia H-1805 A & B, 1934, NYC. - with Gene Sedric's orchestra:
   "I'm so lonesome in the evening, so lonesome all night long
   My man is gone to Chicago, Lawd my man's done gone,

   I'm goin' to Chicago, gonna **catch a rocket plane**,   - - - - -
   ain't got time to ride on no train."

bit Art Works in Santa Fe – made when he was a young man. I tried to use it earlier in the cover of a precursor to "*Cutting Edge*" that I called "*Atom Bomb Stories*". I had hoped the Smithsonian Press would publish it, but a ditzy reviewer who obviously only looked at the cover, wrote me that they don't' "do children's books"! That's why I relegated it to the inside cover.

When I finished writing the book, I started looking for appropriate pictures to better illustrate it. In doing so, I went through hundreds of pictures that were hidden in various places/niches in my PC – a device that I have never learned to conquer (one story herein - in Chapter 8 - demonstrates this chaos nicely). The selection process reminded me of the many good friends I have made in a long lifetime. These include our extended families – my wife and I are both the sole male and female survivors in our generation – and friends we have made in our worlds of music, theater, academia (both attending and professoring), lecturing, sailing, traveling, and engineering. All have made indelible impressions on me and many have been reflected in my writings. I thank them – especially my wife of over 52 years - for their help and inspiration – but mostly for being friends. The majority of those who read this book will know who they are - and will remember me as I remember them.

But, I go out proudly, buoyed up by the wonderful letter of commendation I recently received from the President of the University of Southern California (USC), reproduced on page 124.

PREFACE

THE ENGINEERING LIFE

The eight chapters in this book present stories and opinions garnered from a 60 year career in the burgeoning aerospace world. The stories are supplemental to the ones told in my earlier book, "*On the Cutting Edge*" in that they present background non-technical material about events and people that shaped an important part of the technical world. The progress made during the latter 60 years of the last century, as well as the earlier 30 plus years, is no less than remarkable and has clearly changed the world we live in. The aim of the book is to give the reader some insight into this world, as seen first hand by the author.

"*Rocket Plane*" provides, for the most part, vignettes of actions taken by what is now called the aerospace sector over the last 60 years. Since I have participated as a player in the successive blossomings of the Aviation Age, the Atomic Age, the Guided Missile Age, and the Space Age, I am attempting to provide a kaleidoscope of high points in accomplishments and failures as I encountered them. They cover my pre-engineering life at college at Cornell and in the Navy during WW II; my time in post war grad school at NYU; my work on the atomic bomb in New Mexico; followed successively by jobs in Southern California in the guided missile field at Convair/Pomona and then in space endeavors at Aerojet/Azusa and later at TRW in Redondo Beach, with a stint as a professor at Iowa State University in between; concluded by final retirement from teaching at the University of Southern California, doing some consulting at Microcosm, Inc. and then into the twilight zone, where some interesting consulting and travel adventures took place, and I started writing and publishing books. Being of a semi-technical nature, the stories I tell here are easily understandable for technical and non-technical persons alike. I think they will be fun and enlightening to read. They cover the years 1944 to the present.

In my writing, I depended mostly on my memory, backed by my phenomenal collection of memorabilia. I am a notorious pack rat, and have saved all valuables. My memory is most of the time sensational. I can recall conversations and events from years ago as if they happened yesterday. But, at times I go blank or mushy. In those cases, I have tried to contact those of the participants whom I could find, and asked for their versions. I generally amended the stories as they suggested. But, remember, I'm now an aging geezer trying to do justice to a sometimes funny and always interesting career. And now I leave the arena; undoubtedly as "The World's Oldest Most Unsuccessful Author"!

Chapter 1

THE MAKING OF AN ENGINEER

- PROLOG – GETTING THERE
- CAN YOU STEER, MAN?
- BORN IN TEXAS
- LIKE WEBSTER'S DICTIONARY
- TRANSITION – GRAD SCHOOL FROLICS

GETTING THERE, 1925 – 1946

I must have been 10 years old, when – under the influence of three comic strips; *"Smilin' Jack"* (an avid light plane pilot), *"Buck Rogers"* and *"Flash Gordon"* (the latter two being early Spacemen), I decided I wanted to be an **Aeronautical** Engineer. I'm sure that if there were such a career then, I would have chosen **Aerospace** engineer. I built solid model planes out of balsawood; then took on kits of rubber-band powered models, and finally, with my Father's help in chasing them by car (radio control was not yet available), built gas powered models – culminating in a five foot wingspan Stinson *'Reliant'*. My family and friends encouraged me by giving me wonderful books – circa 1936 vintage – on aviation, which I still retain. So, except for a brief stab at a musical career, I followed my nose, and it wasn't until a mild mid-life crisis in the 80s, that I questioned my career choice – but only for a second.

I was brought up in Philadelphia, having had the luck to be the only son of well-to-do parents. Although they were nominally Jewish, and belonged to a conservative congregation, neither parent took the religious aspects of Judaism seriously. It was no problem for me to go along with this attitude, even though I was properly Bar Mitzvahed at 13, confirmed at 16, and went to Hebrew School two afternoons a week after public school let out. In preparation for Bar Mitzvah, I learned to read Hebrew and still can – it was drummed in so hard – although my vocabulary and understanding is almost zilch. I became an agnostic at 15; an atheist at 17; a position I have maintained ever since and only intend to renege on in my dying moments, as insurance.

This being the 20's and my Mother, being a "flapper", I had a nurse – a wonderful German lady about the same age as my Mother – who greatly formed my character and who lived with us until, at age 16, I departed from Philly to Cornell – never to really return home again. We lived in a row house near Broad and Erie, down the street from Simon Gratz High School, and I went to Grover Cleveland elementary school until I completed second grade. We then moved to a suburb called Germantown and lived in a very large apartment which took up the second floor of what must have been a majestic mansion on Wissahickon Avenue. My nurse and my sister, and I had our own rooms, as did our newly acquired Cook/Maid/Chauffeur couple, William and Marcy. I went to the CW Henry Elementary School, noted for its annual Bird Festival (an avian pageant akin to Swan Lake, wherein a beautiful Cardinal was knocked off by an NRA precursor).

Every summer we spent a month in Atlantic City at a boarding house between the Ambassador and Ritz Carlton hotels. There I learned to love to ocean, and met Al Capone and other hoods and celebrities who intermingled with the society on our stretch of the beach. I went to an historic all-boys High School – Central High – and came into contact with girls only via a dancing class, which my best buddies also attended. I was a very good student and enjoyed Math and Physics and French, but not Latin. I worked on the school paper and in school politics and was fair at the pole vault and good at tennis. Starting at 15, my buddy, Dick, and I used to hitch –hike to and from New York on the week ends to hear the Traditional ('Hot') Jazz music at Nicks in the Village and on 52nd St. One Sunday, on returning, we found a lot of police cars in front of our house. William had shot and killed Marcy in a jealous rage. My lawyer Uncle, Brad Brodsky, got him off in 9 months, whence he went to work at my Father's plant. I practiced my horn playing, on a beautiful cornet purchased from a Mummer, in the basement for about an hour a day.

My Father and I were listening to the radio on Sunday morning, December 7, 1941, when FDR broke in. I was 3 weeks from high school graduation, and had been accepted at Cornell and MIT. Having 'skipped' three half years of elementary school, I started college at age 16. Although I enjoyed playing Hot Jazz, I also wanted to be an Aeronautical Engineer. I chose Mechanical Engineering at Cornell.

CAN YOU STEER, MAN?

Cornell was now on a war-time basis of three semesters per year. I was taking Mechanical Engineering in the Sibley School, was Sports Editor of the thin war-time *CORNELL DAILY SUN*, and still had time to take flying lessons and make solo flights. I was now 18 and had almost completed my junior year when the Philadelphia Selective Service Board sent me its first warning letter. They said they doubted that my student deferment from military service would be continued after the completion of the next semester. It was well known that if you were passive - did nothing - you would probably end up in the infantry. So, when the draft looked inevitable, I signed up in the Navy V-5 program, which was designed to get you ready for training to join the fleet as a carrier pilot, while you continued in School. I loved to fly! For me, the war had started.

For the third time on that fateful day in May 1944, as I pulled back on the stick and cut the throttle, I said to myself, "I'm gonna run this sum'bitch right into the ground!" But, just as on the two previous tries that day - one in the morning, the other earlier in the early afternoon- I had leveled out about 10 feet above the runway and, as before, mushed down with a "plop", followed by a few inevitable bounces. Finally, the indestructible advanced trainer, the renowned STEARMAN biplane, riding on its equally indestructible wide wheel base landing gear, settled down so that I could apply the brakes and begin my taxi back to the hangar area.

As I moved along the tarmac towards my waiting instructor, I reviewed how I had gotten myself in this critical "three times and out" situation at the end of the spring semester. Earlier in the month, as a recent inductee into the Navy V-5 flight cadet program at Cornell University, I had successfully completed basic flight school by soloing in a PIPER J-3 Cub. Like the STEARMAN, it was also a tail dragger and, although I never was able to make a three point landing, I usually "landed" only a little bit high. Instructor Dan had warned me that such high landings could result in ground loops, but my Cub mistakes did not cause him a lot of concern.

To get into the Navy's flight training program, I had had to improve my marginal eyesight and wean myself from my glasses and heretofore myopic lifestyle. I accomplished this by working with eye charts for an hour a day for the best part of the previous six months. I passed the eye test by squinting on the hard parts. My depth perception had checked out O.K., too, so I never did learn what quirk of fate kept me from slicking it down on the runway. It may have been the shear fear of a 'hard' landing or it might have been a simple lack of intestinal fortitude. We'll never know.

The ticklish landing situation was only exacerbated when I moved into the STEARMAN. Like the PIPER Cub, it had a tandem student/instructor set-up, but with separate open cockpits. But, either in the front or back seat, it was very difficult to gage ground distance in the landing attitude. All you could see looking forward was the big engine's top cylinders. If you looked abeam, you could see the ground about a hundred or so feet to the right or left- and this was the datum you had to use to judge your distance above the runway. Referencing the scenery to the side, I made acceptable touch downs about half the time. The other times were bounce-a-thons, but I always maintained directional control. However, the rest of my flying and ground school-learned techniques were fine. At the appropriate time, Dan said, "Ski", you're as ready as you're ever going to be. Let's solo tomorrow, and get you the hell out of secondary school and into the real Navy." The die was cast.

When I reached the staging area and climbed out, I knew from the look on his face that he was mad as hell. "Ski", he said, "You're a menace to yourself and the airplane. You'll never land safely on a carrier! For your own good, and the Navy's, I've got to bust you out." My world immediately turned to ashes. At age 18, my life was over! For the first time in my life, I had failed. My daydreams of shooting down the valiant enemy, white scarf around my neck and a lovely lady panting to give me my reward, dissolved to be immediately replaced by hapless visions of marching in the sweltering heat at the Great Lakes Naval Training Center boot camp. A few weeks later, at the end of the semester, this truly was my new destination and sordid fate.

Years later, two events occurred which made me realize how lucky I was. I attended a meeting of the Aerospace Department Chairman's Association in Corpus Christi, where a visit to the aircraft carrier Lexington, then conducting landing exercises with cadets flying out of Pensacola Naval Air Station, was arranged. We were flown out to the carrier by a twin-engined Grumman COD (Carrier-Onboard-Delivery) aircraft. On the downwind leg, I got a look at the carrier we were going to land on. I had expected to see a huge landing platform. Instead, to my chagrin, I saw a mere postage stamp! Just before we hit the deck, the pilot gunned the engines to assure a successful take-off if the hook didn't engage. What followed was a controlled crash landing! The resultant "snatch", when we a decelerated threw us violently back into our rearward facing seats. I decided then and there that landing on a carrier was the absolute height of human folly.

A later meeting of the same group took place at Mississippi State University, a hot bed of research in gliders. As part of the meeting agenda, we were taken up in a two-person glider which was towed off the ground by a then-ancient but still airworthy STEARMAN. After my glider flight, I asked the tow pilot, an engineering Grad student, if I could ride in the back seat while he towed a colleague up for a turn. I observed his landing technique carefully. Once he had established his final flight path, the sucker actually stood straight up in his cockpit and leaned over the side, thereby getting a perfect look at the ground beneath him! Why hadn't Dan taught me that? Why hadn't I thought of it?

On the other hand, I lived through the war with only a single injured toe incident, which, alas, did not warrant a Purple Heart. So, now, in my dotage, I think that busting out of V-5 may have been the best thing that ever happened to me; at least from a longevity standpoint.

THE MAKING OF AN ENGINEER

THE 'YELLOW PERIL' – THE DREADED STEARMAN BIPLANE STANDARD NAVY TRAINER. THE PILOT SAT IN THE FRONT SEAT; THE STUDENT IN THE REAR. IN THE NOSE-UP LANDING ATTITUDE, YOU SAW NOTHING FORWARD BUT THE TOP CYLINDERS OF THE RADIAL ENGINE. YOU HAD TO GAGE ALTITUDE ABOVE GROUND BY LOOKING WAY OVER TO THE SIDE.

BORN IN TEXAS

After I busted out of secondary flight school, the Navy was faced with the prospect of losing a presumably warm body to the civilian world. I was immediately urged to take a long exam (the Eddy test) to qualify as a Radio Technician. The alternative was to expect a draft notice and become a foot soldier. I chose the outcome of having a bed to sleep in every night and having passed the test, was inducted into the seagoing navy, as a Seaman First Class (S1/c-RT). Thus, at the completion of my junior year semester, they put me on the first train heading west from Oneida to Chicago. There, in the mid-West where I had never before ventured, boot camp and a subsequent year-plus training program to make me a radar-sonar repairmen awaited.

I first met the gang from Texas on a sun-lit railroad station quay in downtown Chicago. It was in the late spring of 1944 and we were all waiting for a Navy-commandeered local to take us a bit north to the Great Lakes Naval Training Center where we were to undergo a 6 week intensive boot camp course before being unleashed on the unsuspecting evil Axis forces. I was still in a state of shock, my life having been completely destroyed a few weeks ago.

Then, I was a relatively carefree junior year Cornell University engineering college student fighting the war as a Navy V5 Flight Program cadet. My goal had been to graduate from secondary flight training school and go on to Midshipman training in the Naval Air Force. Now, busted out because on my final check-out flight I wasn't able to safely land the storied Stearman biplane trainer, my heroic aspirations were dashed and my normally optimistic mien had turned forlorn. At age 18, my life was essentially over! For the first time, I had failed at something I dearly wanted to accomplish. My dreams of becoming and officer and serving on a magnificent aircraft carrier were shattered. Only the indomitable will to live kept me going.

I was a loner on the quay, most of the other inductees having arrived as a group from various outlying middle-west, southern, and eastern states. I was attracted to the Texans, lounging on their ditty bags, because of their unmistakable drawls, their uniformly bright looking and clean appearance, their obvious joie de vivre and camaraderie, and their predilection, armed with ukes and guitars and wonderfully harmonious and nasal twanged voices, to break out into song. The songs they sang were then called "hill-billy" songs made popular in the south by Roy Acuff and other country/western pioneers. I put my ditty bag down near them and hummed along, knowing some – but loving all of the words they zestfully sang.

During the semi-rigorous 6 weeks of boot camp we became good friends. They adopted me as they would a stray dogie. The Texans were all engineering sophomores or juniors from Rice Institute, a selective school in Houston that awarded free tuition to outstanding Texan student applicants. I often think they regarded me as a novelty - being both a Yankee and probably the first Jew that they had ever been close to. It soon became apparent that we had a further bond in addition to the love of music. None of us (there were two exceptions) were very zealous or serious about the ongoing World War, nor in any way concerned about its possible future dangers. We were too close to college to be infuriated by such things as Nazis and Samurais. We equally hated authority in any of its manifestations, all of us being spoiled middle class brats. These attitudes bonded us to consider the Chiefs and other G.I. non-coms in charge of our training as the real enemy. While we were busy earning our "Zone of the Interior" merit medals, the real war was very far away, while the promise of our eventual first liberty in Chicago very near and desirable.

"Sthuart" Sinclair promptly dubbed me "Broadski" in his endearing high whining voice (his was one of the high - harmony voices) that also had a tinge of a lisp to it. This sobriquet stuck to me throughout the boot camp experience; later - after we separated and joined the "real' navy- to segue into "Ski"- the inevitable Navy name given to anyone whose last name ended in 'sky' or 'ski'. I immediately took a liking to Stuart. He was small and obviously very bright, and had a nice laissez-faire outlook on the world. Like me, he was a natural born fuck-up as far as naval things were concerned, and we usually managed to find a secluded nook to sit around and talk while the rest of the bunch were marching or doing other necessary training maneuvers. Right after the war, I visited him at his home in Houston. I found that he and his father had a daily hobby of making and recording paper bets on horse races at ten tracks around the country and keeping an account - in several rows of filing cabinets - of their results. By now, they were millionaires on paper. But, they would never bet real money. "That would take the fun out of it", Sthuart said.

'STHUART' IN BOOT CAMP – 1944

Ewell (pronounced 'Yule') Clarke was the acknowledged leader of the group. He was tall, handsome, and had a very mellifluous voice which made him the lead singer. He sort of watched over the group like a mother hen and was the arbitrator in the infrequent squabbles that broke out among them. He ruled by threatening to "tell your Mother" if things got out of hand. Of the others in the close knit group, which included Walt Hearn with his ever present cowboy boots, the Simonds brothers, Robert and Floyd, Algy Badger, Tinky Manry, only Dr. Jim Wilhoit remains as a friend for life.; though I do stay in touch with Dr. Algy. "Jimhoit" (all the gang had nicknames) was Stuart's friend and we thus gravitated together.

Jim was also a lead singer and had the genre down to a tee. I tried to copy his nasal twang, but usually ended off key, which no one seemed to mind since it apparently added some authenticity. After the war, Jimhoit got his Doctor's degree in Engineering Mechanics at Stanford, worked in industry at Convair/Ft Worth and ended up as a professor at Rice and Texas A&M. In late middle life, he inherited a large tobacco and cattle farm in Kentucky. In 2000, he turned over the spread to his kids and moved into nearby Versailles (pronounced Ver-salles) and promises us a wonderful visit if we ever get there.

The six weeks of boot camp went by quickly. Previously such training had been almost twice as long, but the pressures of war were causing an accelerated demand for new bodies to join the fleet. We did a lot of drilling, which did not make much sense since sailors hardly ever walk around carrying rifles. We learned to fire what appeared to be WW I Springfield rifles from the prone position. I used my left elbow as a fulcrum and waited for the aim sight notch of the very heavy and unwieldy weapon to approach the target center as the rifle barrel peregrinated back and forth in a plane parallel to the ground. Then, I squeezed the trigger and braced for the considerable recoil. We got to shoot ten real rounds. On ex-post facto examination, my target had 9 holes in it; my next door neighbor's had eleven. So much for "Dead Eye Dick"!

EWELL (LEFT) AND JIMHOIT (RIGHT) – 1946-ISH POSTWAR

When we weren't marching or making hospital corners on our cot's bedclothes or doing pushups or swabbing the decks, we did learn some Navy lore and knot tying methods. There were periodic breaks from the tedium, and during these, at the tender age of 18, I first started to smoke real cigarettes. Previously I had had some experience with smoking marijuana with my musician friends. At boot camp, when a break was called, the guys would assemble at a rest area and immediately light up. I couldn't resist the apparent comradeship invoked by the habit and so embarked on the evil curse that would not be broken for over thirty years. The temptation was simply too great to resist. I soon got hooked and then started to buy my own ("Lucky Strike green has gone to War"). On the plus side, however, it better prepared me to learn my limitations when accepting Mary Jane bounties that my musician friends and musical groupies offered me later on. Albeit still a virgin, I was slowly being inducted into the adult world.

After such breaks, Stuart and Jimhoit and I would often seek a secluded area and hide out, if we thought we could get away with missing a drill exercise. We knew of storage areas where we could relax out of view of casual passers-by. We never got caught, though I think we were missed on occasion. If asked, we used "sick bay" as an excuse for our absence. Later events, having to do with future duty assignments make me suspect that somewhere along the line, an asterisk was placed after my name with a "fuck-off" connotation.

On the weekends, we had to dress formally in our whites with the brown puttees wrapped around our lower legs - the mark of a "boot"- and do formal marching reviews before the powers-that-be. On Sunday mornings, attending services in the chapel was probably mandatory but, as I recall, allowed time to drown out the chaplain's bombast with thoughts of home and school and 'whither thou go'est' or grab a cat nap.

One of the negative highlights of boot camp occurred toward the end. Part of our training included passing some rather non-rigorous, at least for me, swimming tests. However, the denouement of this saving-your-life-at-sea training was the death defying forty(?) foot jump (in a 1998 correspondence, Jimhoit said, "It just seemed like forty!") into a pool, fully clothed, with subsequent clothing removal and swim "to the shore" after entry. Had it not been rigged so that our group ascended the ladder to the dive platform one after the other, with a Chief continually yelling "keep it moving, swabbies!", I doubt if I would have had the courage to make the jump. At the brink, I waited for the guy ahead of me to clear out and, with my eyes closed, abandoned all hope for what seemed an eternity. As instructed, I hit the water with my feet and soon was scrabbling to get out of the way of the next jumper. With this deed accomplished, the entire boot camp ordeal was essentially over.

The day of the graduation ceremony (puttees having been returned to Stores as our last act of contrition) our whole company was disbanded and reassigned to our new duty stations. My close association with the Texans ended there - to be renewed after the war. I was unceremoniously loaded onto a train that took us to the next one-month stop on the road to radio technician Valhalla: the pristine white armory at Michigan City, Indiana, roughly on the eastern side of the lake opposite Chicago. Many of the others were shipped off to "pre-radio" program venues at junior colleges in the greater Chicago area- but the screw-ups like me were banished, sans leave, to the hinterlands.

Here we had a routine of about six hours of class time intermixed with drill and fun and games on the adjacent beach. On our first weekend, we were finally granted leave at noon Saturday, after review, and Sunday after services. The catch was that we had to report back to the barracks by 8 pm, making a journey to the big city an impracticality. We sadly noted that women did not appear anywhere near our bit of beach, probably having been warned of the pent-up appetites of the trainees. But, by and large, this was a nice sojourn. The classes reviewed electrical engineering and math basics which, in spite of my rather cavalier just-get-by student attitude in my pre-Navy years at Cornell, gave me no problem. The days went quickly.

After the first acclimatization weekend, I spent Saturday afternoons going by bus to a nearby light plane airport which had a single lake-facing grass runway which sort of sagged in the middle as you took off. About half way down the runway you lost sight of the horizon until you rolled back up the hill. They had 45 and 65 horsepower Taylor Cubs for rent. The latter version was very similar to the Piper Cub in which I had learned to fly in basic training earlier, except it had side-by-side seating for two passengers, instead of the tandem seating arrangement (instructor up forward) of the Piper. Solo time was $2 an hour; $3 for dual. After a takeoff and landing check out by an instructor, they allowed me to solo, after informing me of the local airway restrictions.

My worst experience occurred in my second - out of three - flight weekends. Without knowing it, I drew the underpowered Cub on a day when there was a fairly stiff breeze off of the lake. Take-off was uneventful until, just beyond the end of the runway at about 150 feet altitude, I ran into a steady wind current whose velocity was just about the same as the plane's indicated air speed. I just seemed to hang in mid air, making no forward progress. As the plane slowly gained altitude, I realized that the current was very thick and would keep me gaining altitude. I began to sweat in this new situation. "What the hell should I do", I wondered. Usually by this time in a solo flight I would be happily singing to myself at the top of my voice, as I was wont to do. Flying solo is normally an ethereal experience, giving you the feeling of the freedom of the birds. Now I was fearing for my very life. I was afraid to turn lest I would lose enough altitude to hit the ground. After the initial panic subsided, I decided to wait until I rose to a thousand feet and then make the turn. This worked wonderfully and I immediately made a quick entry into the landing pattern and got back to terra firma just as quickly as I could. On the ground, I let out one big sigh of relief. The next Saturday, I made sure I rented the 65 horsepower Cub.

Toward the end of out training month, we were asked to list, in order of our preference, the first three sites for our forthcoming three month primary school and the subsequent, and final, nine month Secondary radio / radar / sonar technician school. The site choices for primary school were either Monterey in California; returning to Great Lakes; the University of Houston; or Gulfport, Mississippi. Naturally, Monterey was everyone's first choice and Gulfport their last. My first choice for secondary school was Yerba Buena (or, Treasure) Island in San Francisco Bay, a half - way stop along the Oakland Bay Bridge, since the Lu Watters band, with Turk Murphy, and several other great jazz bands were playing in the vicinity. My second choice, Navy Pier just off the Loop in downtown Chicago, was also a "no brainer", since the big city had a wonderful reputation for providing servicemen with great hospitality and I knew there was great music to be heard along Rush Street and on the South Side. The Texans, of course, opted for Corpus Christi or Treasure Island for secondary school, and lucky Jimhoit and Stuart landed at the latter. To my chagrin, I drew the Gulfport primary school short straw, but got Navy Pier for secondary as a partial recompense.

At pre-radio graduation, I got a 10 day leave and a train ticket to Philadelphia, with orders to board a train to Mississippi from the Reading Terminal twelve days later. So, I sadly said goodbye to the Texans who had greatly helped me ease the way from college to a sort of "real world" of widespread war. I saw Stuart and Jimhoit immediately after our discharge in 1946 and keep current with both to this day. After the war, I was hosted by Jim's brother, Bill, in El Cerrito just north of Berkeley, where I went to hear Turk Murphy. The great man was playing with gusty abandon at a saloon he partially owned there, called "Hambone Kelly's".

THE MAKING OF AN ENGINEER

I remember all the Texans fondly, for they were a gentle down-to-earth bunch who had a certain elan and took much pleasure in life. One of their number, on the edge of the clique, asked me if I wanted to be saved; another, not in the clique, made a homosexual proposition to me as we stood at adjacent urinals. I politely refused both offers. The songs the Texans sang may be seen and heard in my book, "*Songs My Mother Never Sang To Me*" (Amazon.com/books)

On my arrival back home for leave in Philly, I was treated like a returning war hero. Little did they know!

LIKE WEBSTER'S DICTIONARY

My Navy career was variegated after I graduated from Boot Camp at the Great Lakes Naval Training Center, north of Chicago. I was in the Radio Technician program, learning how to maintain and repair radar and sonar units. School took me to Michigan City, Indiana; Gulfport, Mississippi; back to good old Chicago for 9 glorious months; thence to Link Trainer (blind flying instruction device) school in Atlanta; and most recently to Brunswick, Maine where I plied the Link Trainer Instructor trade and occasionally flew in a blimp looking for U-Boats and earning flight pay. Now, in the spring of 1945, I was fighting the war on Cape Cod and suddenly found myself, as in the Bob Hope / Bing Crosby epic "*Road to Morocco*" title song, '**Like Webster's Dictionary, I'M MOROCCO BOUND**'!

"We're shoving off the day after tomorrow on MATS (Military Air Transport Service)", Lt. Sullivan, the base training officer, said, and began making assignments to all of the seamen he was nurse-maiding. When he got to me, "Ski, you're going to run Link Trainer flight simulation classes on-board, and chip paint when you have no pilot training scheduled. You'll set up the new version Link that we'll bring with us and deep six their old one." At the end, he said we would all get overseas pay, an overseas medal to hang on our middies, and some liberty time in Rabat. He estimated it would take us 2-3 weeks to get the war-battered carrier USS Valley Forge, now laid up in Port Lyautey, refurbished to resume military action.

After almost 2 years in the Navy, serving mostly in the dangerous waters of Gulfport, Mississippi; Chicago, Illinois; Brunswick, Maine; and Falmouth, Mass., this would be my first overseas billet. I was excited! Oh, boy! A chance to go where the real action was, and to have liberty in a foreign country! Join the Navy and see the world, and perhaps even cherchez la foreign femme!

I 'fought' five months of WWII in the spring/summer of 1945 on assignment to CASU (Carrier Aircraft Service Unit) 26 at NAF Otis Field at the far end of the huge Army training base, Camp Edwards, in Falmouth, Mass., at the southwestern end of Cape Cod. CASU units had dual functions: to train and/or re-equip naval aviators needing R&R or being rotated from service in the Atlantic to the Pacific or, in a rapid reaction mode, to be shipped as a total unit to the site of a carrier laid up because of war wounds or malfunctions, to carry on repair and training functions. Included in a CASU's cadre were experts on all things nautical. My first and only 'call to overseas action' came about halfway through my Otis Field assignment.

I had two distinct jobs on the Cape: One, as a Link Trainer Instructor one week per month and the other as a F4U Plane Captain on the flight line the rest of the time. Both jobs were fun; neither was strenuous and, if I was careful and lucky, nor was life threatening, except when I started the F4U's engine every morning. My plane was driven by Commander Tommy Blackburn, an Admiral's son, later to become a hero of the war in the Pacific. He was taciturn, but a very nice guy. In a kidding way, he never missed an opportunity to razz me about busting out of flight school and now 'flying' a toy airplane like the Link Trainer.

The engine starting procedure for the earlier F4U fighters consisted of inserting a 3/4" diameter, 6" long black powder cartridge into a breech located on the instrument panel to the right of the pilot's seat. After locking the breech, I pulled back on a spring-loaded firing pin, and let go. In theory, this would ignite the powder, which would drive a turbine which in turn revved up a flywheel. When the flywheel reached operating speed, it would engage a gear train which would then turn the propeller. The problem was that I could anticipate, but never expect, a back-fire two to three times a week. When this occurred, the breech flew open with a bang and the cockpit was explosively filled with sooty black smoke and I was transformed into a minstrel. Worse, despite knowing it could happen, it always scared the shit out of me. It took almost an hour to clean up the mess. But, the plane captain's job, whose hours were from 4 a.m. to 11 a.m., not only earned me the right to print my name on my aircraft below the pilot's name but also allowed me time to go sailing. I cheerfully did this almost every afternoon on Buzzard's Bay in the 21 foot Star boat I had bought with two other flight line buddies for $500. So, I just grinned at the backfires and wrote them off as some of the unbearable fortunes of war.

Prior to the CASU 26 assignment, I had spent three months at NAF Buckhead, in the Atlanta suburbs, going to and then teaching at Link Trainer instructor's school. The Link was a rudimentary flight simulator which helped train Navy pilots and maintained their proficiency. The teaching gig assignment patriotically allowed me to replace a WAVE friend, Mary Margaret Maureen McMichael of Des Moines, for overseas (Hawaii) duty. When I graduated, I was a Seaman First Class, Specialist T (for teacher), LT (for Link Trainer). The purpose of these flight simulators was to familiarize pilots with blind and night flying and navigational techniques, and to acquaint them with the proper radio language for communications with ground stations or their aircraft carriers. Most important, it showed them how to find their carrier returning from a mission.

In appearance, they looked like stubby low wing monoplanes, replete with short wings and small tail surfaces, and painted in the same yellow color scheme of the dreaded Stearman biplanes, in which all my students had flown in advanced flight school. Since the latter was the plane that had caused me to bust out of flight training at the beginning of my illustrious naval career, I got heart burn every time I looked at this reminder of my former nemesis.

The Links, sitting as they did in the middle of the classroom, were wondrously simple. The cockpit, entered by climbing a short ladder, was covered with a big hinged hood, sealed tight for leaks, and was lit with a low wattage over-head light mounted in the hood ceiling. The instruments on the panel were lit, and were few and very basic. They showed airspeed, altitude, rate of climb, and engine RPM and temperature. Aside from a compass and a radio set, the main instrument for blind flying assistance was a combination artificial horizon, which indicated whether you were climbing or diving, with a bank-and-turn indicator, which indicated whether the wings were parallel to the horizon or whether you were in a turn.

The Link Trainer could rotate in a circle in response to rudder pedal inputs, while a bellows-like arrangement allowed limited pitch-up and -down and roll right and left motion.

The earlier versions remained horizontally fixed: no rolling or pitching. The artificial horizon/ bank-and-turn instrument would respond properly to inputs from the control 'stick'. As an added harassment, an evil instructor could turn on "rough air" by the twist of a valve. The resultant 'turbulence' would result in the trainer bouncing a bit, if you cared to notice this bit of hardly stark realism.

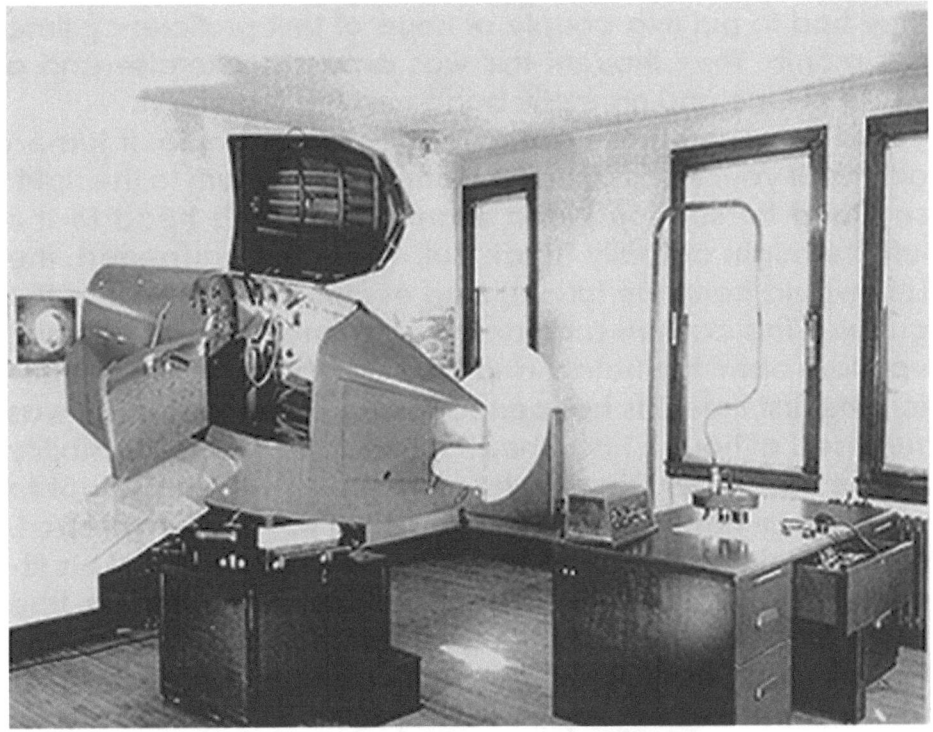

PHOTO. COURTESY OF SPECIAL COLLECTIONS, BINGHAMTON UNIV.

THE LINK TRAINER IN ALL ITS GLORY! NOTE THE 'CRAB' ON THE DESK. NAVY LINKs WERE BRIGHT 'STEARMAN' YELLOW

While the pilot trainee "flew" the trainer, the instructor sat at a desk with airways charts on it; these charts being exact copies of the charts the pilot was using in the ongoing training exercise. A wheeled black-erasable-ink writing device, called a "crab', rolled over a chart on the desk plotting the course the pilot was navigating. A hot shot instructor like me could 'draw' perfect circles on the chart when 'flying' the Link in a constant altitude turn. The instructor also had earphones and a mike to play the part of the base station operator for the student pilot. A typical training session lasted about a 40 minutes, and was uniformly hated by all the real pilots. By decree, they had to put in a couple of hours of Link proficiency time per month. They thought this was a puerile exercise and a waste of time and normally badgered the Instructors.

But, we sometimes got revenge. For example, it turned out that if your student was making a steady turn to the right, say, and he suddenly and sharply turned his head to the left, he might actually black out. When this happened, the Link would continue turning and eventually 'crash' into the ground. This caused an alarm to ring and the pilot to wake up, slam open the hatch, and look sheepishly at the instructor. The first time this happened to one of my students, I was surprised although I had been forewarned of the possibility by the old-timers. The hapless Top Gun begged me not to report what happened. "Ski", you'll not put this in my report, right?" I solemnly agreed to leave the incident out of his fitness report, and I really did not write it up. Nevertheless, true to the tradition, word of the 'crash' would somehow get out verbally, and the pilot would take a terrible hazing from his fellow pilots and nearby plane captains.

We landed in Morocco at an airstrip near the port, and were assigned bunks on the Valley Forge, now having a skeleton crew during repairs and immediately told to get busy. What, no liberty after the long plane ride? No, "Liberty when our work is done", said Lt. Sullivan, himself mad about his equally depriving orders. Most of Ship's Company, having undergone a vicious attack, were on leave during repairs. Recruiting a few sad sack left-behinds, I uncrated and set up the new Link in the ship's training room, and established a training schedule for the pilots who were now operating their planes off the nearby airstrip. When I wasn't instructing, I was, just as Sully predicted, chipping or painting. We worked our asses off in ten hour days for over two weeks.

When the job was finally done, we were given a late afternoon / evening liberty. That was it! We were to be flown back to the Cape the very next day. Well, at least we would get one night on the town. I bonded with three of my fellow plane captain-paint chippers and we took a cab into Rabat. We sight-saw around a bit, found no women, and decided we were hungry. We tried what looked like a nice restaurant, and tried local food and drink. It was different, but good. However, I felt queasy almost the minute after I left the restaurant. I was having a massive ptomaine attack. I soon threw up in the street. My stomach and head ached, and I was very woozy. I simply couldn't go on.

One of my buddies, who was also having minor cramps and had had enough of the night life, bundled me into a cab and helped me to sick bay. I stayed there. miserable for three days, while the rest of my CASU went home. When I was finally discharged, they put me on a plane to Norfolk, and I then found my way back to my unit by train and bus.

Such was my first, and only, overseas adventure in WWII. My contribution to the war effort consisted of chipping paint, teaching a few pilots, and suffering Ptomaine.

However, I did get more overseas pay than the others because of the extra time I had spent on foreign soil. Fortunately, over twenty years later, I had the opportunity to revisit Rabat under more favorable circumstances. This time, I scored a genuine Moroccan rug instead of Ptomaine. This adventure is covered in detail in a story, "The Road to Maroc" in my book, "On the Cutting Edge".

CATCH A ROCKET PLANE

GRAD SCHOOL FROLICS (1947-50)

I was mustered out over two years later in May 1946, no worse for wear and tear. Although I was generally a screw up in the service, I did nothing to harm our war effort. I literally grew up during my time in the Navy; from a wise guy kid to a wise guy adult with a small touch of maturity. I got an 'early' discharge for the ostensible reason of resuming my college education in June, which I actually did resume in the Fall. My discharge was expedited by my taking a wild chance by telling a yeoman who ran discharges that I was indeed a cousin of his good friend Joe Brodsky of Newark, who was to be married four days hence. Of course, I had never heard of Joe, and knew of no family branch in New Jersey. Thinking quickly, I told his friend the Yeoman that I really wanted to go to Cousin Joe's wedding. "You got it, Sailor!" I was out in two days!

Out of the Navy, I gravitated to New York and tried to get back into the hot jazz life. My Philadelphia buddies had wangled their way into a six night a week gig in Greenwich Village. Pay was $25 a week plus all you could eat and drink. I sat in with them until it was time to return to Cornell to finish my senior year. I tried to get down to the city on the weekends. Despite this diversion, and writing a humor column for the *Cornell Daily Sun*, I got very good grades and graduated early in May, 1947.

I was at loose ends about a career. I moved back to the Village, 108 W. 12th St, and played regularly. As Fall approached, I had to decide whether to go to graduate school or really try to make it as a jazz musician. On one hand, New York University had a degree program in aeronautical engineering, which field had always attracted me. I still had the GI Bill money available and I could make a deal for additional support proof reading for a famous

Professor who was writing an aeronautical encyclopedia. On the other hand, my cornet playing sometimes bugged my compatriots. It seemed that I played a quarter tone sharp, and could not hear it. But, I loved the life and the music. I had resumed taking lessons from Frankie Newton, a jazz giant. Perhaps I could do both? Fate, in the person of Johnny Windhurst, a very lyrical though much more subdued, cornet player came along and settled the issue. He always played on key. He played on the weekend; I played weekdays. The great majority of the GI's who had returned to college had one sole objective: Get it out of the way and get on with lives that had been interrupted by the war. So, when I returned to Cornell in the Fall of '46 to complete my senior year, my attitude was no longer that of just trying to get by. I was a bear for knowledge and got straight A's in spite of running down to Greenwich Village on weekends to hear or play the music. My newly acquired ability to get by on 3-4 hours of sleep a night stood me in good stead.

After graduation and a permanent move to the Village, life became a bit more complicated. I played on Sullivan Street most of the summer, and continued this 9 p.m. to 4 a.m. routine after grad school started in the Fall. In the first year of grad school, I was a full time student, taking courses from about 4 in the afternoon 'til 7 or 8 at night, generally at the downtown Washington Square campus near my room. During the early morning, I slept (after our gigs, we generally tapered off by going to my room and talking, playing my fabulous 78 rpm records, drinking beer and smoking until around 6 a.m.) for a while, then did homework – which continued on the IRT to the uptown engineering campus in the Bronx. There I proof read the encyclopedia that my mentor was writing at 10 cents per page pay, occasionally having an early class there, too.

In my second and third years, major life style changes took place. I was hired by the Mechanical Engineering Department as an Instructor. I began teaching two classes; one a lecture, the other a laboratory class. I now had my first Master's degree and was going for the Doctor's. I played only on rare weekday occasions when the kid who ousted me couldn't make it. The teaching was a blast! The students were all so avid and so questioning that this must surely have been the golden age of teaching. During this period, I found time to get married and have a baby daughter. We moved uptown to more sedate surroundings. I was settling down to a life in engineering and academia; left with the question I still have: Could I have made in the world of the old time funky jazz that I continue to love so well – now along with only a world's handful of fellow Moldy Figs?

The three years at NYU went by quickly And 1949 was coming to a close. As completion of my Doctoral thesis neared, I started looking for gainful employment. As you will see, fate – in the form of flying saucer visions – landed me in the Wild West; Albuquerque, New Mexico, no less!

Chapter 2

ATOM BOMB STORIES

- **THERE'LL BE SOME CHANGES MADE**
- **ALBUQUERQUE – 1950**
- **WHAT! NO FLYING SAUCERS?**
- **SETTLING IN AT SANDIA**
- **BECOMING LANDED GENTRY**
- **A PAEAN TO ALAN POPE**
- **SOMETIMES A GREAT NOTION**
- **TURNED OFF AND GONE**

THERE'LL BE SOME CHANGES MADE

After the war, the defense industry, long the bulwark and vanguard of our technical excellence, was suffering greatly. There was a glut of war machines, many moldering in storage in the Arizona desert. The pinch was felt most by the gargantuan aeronautical-based industry. They had no new military orders and few prospects in the commercial arena. G.I. returnees who took course work on aeronautics were looked on as hopeless romantics. Nevertheless, having known that I wanted to be an aeronautical engineer since age 10, and realizing that I would never make it big as a hot jazz musician, I knew what I had to do.

As the end of the 40's approached, a subtle change pervaded the national psyche. The cold war realization began to settle in, and a powder keg atmosphere started to develop. Unbeknownst to the public, the US was forging into the atomic age, and the Russians were frantically attempting to do likewise. The Korean war rumblings were beginning to be felt and Ike's "military-industrial complex", timidly but resolutely, again began to flex its mighty muscles. By 1950, the beginning of the next inevitable 20 year boom and bust cycle which has been uniquely peculiar to the U.S. aerospace industry was at hand. The era of John Foster Dulles' Massive Retaliation protocol was fast approaching. *"Bring out the Bugles, boys; We'll sing another song!"*

Almost at the outset of the interview, Pete Petersen asked, "Do you have any problems about working on a highly secret government project? Sorry, I can't tell you any more" My ears immediately stood up, my spine stiffened, I inwardly giggled, I kept a poker face. This was it! I knew it in my bones! I had hit the mother lode! These sucker's were making flying saucers and I was getting in on the ground floor!

By late '48, it became apparent that, barring a calamity, I would finish my doctoral thesis within a year. It was time to start looking for a job. It was a terrible time economically in the aerospace industry. The war was over and production and new development had stopped. Job opportunities were few and far between. It was even worse for my New York City classmates since, by culture and native provinciality, they were precluded from living in the hinterlands west of the Hudson River. Both of the local airplane companies, Grumman and Republic Aviation were now in a complete state of disarray, their next projects seemingly having to come from heaven. It was much easier for me, an already uprooted Philadelphian, to consider a job anywhere.

After considerable correspondence and phone exchanges and an unappealing interview trip to Bell Aircraft in unappealing Buffalo, I found two intriguing job possibilities to which I applied. Both candidate companies agreed to a shared-expense sequential interview trip. One position was with old line airplane builder, Convair, in San Diego;

the other with an unknown company by the name of Sandia Corporation. I had found their blind ad for "Interesting advanced government work in Albuquerque" in the New York Times.

I went to California - my first time west of the Mississippi since my post discharge quick foray into LA and San Fran in my mother's '46 Ford convertible - and immediately liked what I saw. The ambiance was startlingly different from the drab East. The beloved ocean with its balmy sea breeze was a delight. They wanted me to work in the Aerodynamics Group on a water-landing/take-off version of their delta-winged F-102 fighter, and a new commercial Jet that they would soon be starting on. It was very tempting - and I almost said OK right on the spot, taking the huge $425 per month salary and running! After all, my instructor's job at NYU was paying only $3600 per year, albeit augmented by my GI Bill stipend, which was rapidly running out of eligibility, of around $100 per month.

In the back of my mind, however, was the overpowering thought that there might be much more about the work at Sandia than met the already very myopic eye. For almost a year now, reports of almost daily sightings of UFOs in the southwest were flooding the newspapers. In fact, because of their frequency, such reports were already being relegated to other than the front page. "Could this be the Sandia action?", I mused as I flew eastward, and immediately determined to ask penetrating questions before making a pro-CONVAIR decision. The idea of working on flying saucers was just too delicious to allow me any other course of action.

As the airplane approached Albuquerque in the late afternoon, the scenery was like none that I had ever seen before. The vast desert areas culminated by the Sandia - Manzano mountain ranges were starkly beautiful. Sandia Mountain itself was a soft reddish color. I was later told that

Sandia means "watermelon" in Spanish - a translation that I have never seriously checked to this day because of its inherent charm.

The Rio Grande valley was green with spring foliage supplementing the fir tree varieties, all in stark contrast with the desert back ground. The town was then almost 40,000 in population, with the only paved roads being to the airport and in the downtown area out to the University. Central Avenue - Route 66 - also paved, windingly led from the west valley, across the river and through the small downtown area towards the distant Tijeras Canyon pass that separated the Sandia's from the Manzano's.

Sandia Base, the home of Sandia Corp., was in the southeast quadrant of the now-sprawling city, close to Central Avenue at the foot of the Manzano range. The cab driver told me that Sandia Corporation was only a few years old and now employed around 700 people and was growing rapidly. The civilian part of the Base consisted of a spate of new construction being built on land ceded by the Air Force to the Atomic Energy Commission. Many Sandians, attracted by the available low rental houses formerly occupied by military personnel, had moved into former G.I. base housing. They were now building an Employees Club, a civilian version of the nearby Kirtland AFB's Air Force Officers Club. It would be replete with swimming pools and restaurants. This was only one of the many new facilities under construction in anticipation of a large expansion of the Sandia plant. I was told that such amenities were the showcase part of the ongoing campaign to recruit new personnel, since it was difficult to get people to move to what was then viewed as an overgrown cavalry outpost. Recruitment was made even more difficult because of the very non-job specific come-ons such as the one I had read in the *Times*, and the complete lack of public knowledge of the reason for their existence.

The meat and potatoes job interview was pleasant socially, but not very enlightening technically. I met with the Director of Research, Robert P. (call me 'Pete') Petersen, a large immaculately dressed man with a somewhat florid complexion. He was obviously a westerner by aura, a Californian by birth, and immediately likeable. The interview took place on HIS territory, the West Lab site on Gibson Avenue across the street from Kirtland AFB (which continues to double as the civil aviation field). West Lab had been a swank private girl's school, and had lovely facilities and a beautiful campus; alas now surrounded by a heavily guarded two layer security fence. Aside from that, it looked like an ideal place to work, with large private or 2/3 person offices and well kept grounds with fountains and walkways.

Along with Dr. Petersen were his three lieutenants. All were clad in what I was to learn was the "official" Sandia uniform of the day: blue jeans and Hawaiian shirts. Quite a difference from the very formal East. Wes Carnahan was the electronics major-domo; Ev Cox, the science head; and Ken Erickson, the technical staff head, who would soon become my boss, and George Hansche, who would become my early mentor. All had their doctorates, all were physicists, and all had been recruited from the Applied Physics Lab of the Johns Hopkins University. It turned out that they were there because their individual fields of expertise, mostly concerned with proximity fusing (exploding before impact with the target) that they developed during the war; a now critical element in Sandia's mission.

Wes had a withered left arm, which I tried to avoid looking at during the interview. I later found out that he was a leader in the newly developing field of FM/FM telemetry - a radio link method of transmitting a high volume of data over long distances with little signal noise and distortion - and essential to Sandia's need to send encoded instrument readings to a ground station from a falling bomb during weapons tests. He would later leave Sandia to join a new outfit, Varian Brothers, in Palo Alto, to pioneer in the development of the Klystron tube. This high power electron tube, in some applications a very large version of a radio vacuum tube, permitted the telemetry world to greatly increase the carrier frequencies into billions of cycles per second and paved the way for the present satellite communications revolution.

Dr. Everett Cox appeared to be a gentle philosophical type who also becamea friendly mentor to me. He later left to head up research at RCA Whirlpool, and undoubtedly helped develop a better washer, although I felt that this may have been a waste of his keen analytical talent. Dr. Ken Erickson, like Pete, was over six two, and unmistakably a Texan - even to slightly bowed legs. I was pleasantly taken in by the genuine affability and camaraderie of my interviewers, even though my war time experience in Chicago had already taught me that Easterners may be unique in their built-in standoffishness. I liked these guys almost from the beginning of the feeling-out small talk phase of the interview. As we settled down to cases, I became anxious to hear their sales message.

Things became immediately interesting. Pete said that, unfortunately, because of the secrecy and delicacy of their mission, he could not tell me what they were doing other than "highly secret government work". He then questioned whether I would morally object to working on very Top Secret defense developments. At this, my heart leaped!! I, wise guy that I was, immediately knew exactly what they were doing! I had somehow stumbled on to the source of the flying saucer business - and at a handsome promised $450 per month, 30 day per year vacation rate! From there on in, I don't remember listening to too much input, but rather tried to ingratiate myself. My mind was made up. I did wise off to Wes when he inquired about my interest in working in electronics. I was such a fluid flow analysis nut at that time, having devoted the last three years of my life in such pursuits, that I allowed that I would only deign to turn on light switches! Despite this, I was given a formal verbal offer as I left, which after discussion with wife #1 and a few close friends, I accepted contingent on my successful completion of my doctor's degree work.

It turned out that the dreaded calamity all doctoral candidates fear did occur shortly after the interview trip, but circumstances of new age advanced technology still allowed a finish on schedule. My doctoral thesis completion hinged on getting a reasonable answer following the successful inversion of a 30x30 matrix of long numbers with many decimal places. This was a complex algebraic expression of 30 rows and 30 columns that had to be figuratively turned inside out. I had been working on the calculation for six months, painstakingly using a Marchand mechanical calculator. One day, doing a check, I learned to my utter revulsion that I had early-on made a mistake. After pondering on it a while, I knew that the error it caused would be of only minor importance and that no one but me would ever know about it. I put it out of my head and forged on.

But, the mistake preyed on my mind. It would be dishonest to ignore and it could haunt me forever. I finally damned the torpedoes and told my major professor about it. In his typical up-beat way, he grinned and said, "Not to worry!" He seized on this as an opportunity to arrange for me to go to the city of my nativity where the University of Pennsylvania had begun operating the world's first practical digital computer - the ENIAC. What a historical opportunity for both his reputation and for me! He arranged for me to use the computer over a month's time span-mostly, as it turned out, in the dead of night. The machine was a horror of early electronic design. Racks and racks of vacuum tube filled equipment took up two large primitively cooled rooms.

Technicians, ready to chase troubles and change tubes, stood by around the clock. I lived at my parents, and ate at home. I was trained on the fly at odd hours, and in a week knew how to program the monster.

Miraculously, in three non-consecutive nights of operating, I completed the work that had taken me over six months on the Marchand. The numbers came out reasonably, and I subsequently easily defended my thesis successfully, aided greatly by a vicious drawn out verbal battle between my major professor, Dr C.T. Wang and his great rival on the faculty, Dr J.L. Ludloff. This exchange began immediately after I tried to field the third question that Dr. L, hoping to show up his enemy's student, threw at me. After the battle subsided, they were both so winded as to give short shrift to the questions they had for me and said that I had passed my thesis defense with flying colors! Now you know why they call the PhD, "Piled Higher and Deeper".

And so, the young Brodsky family - with wife #1 and baby daughter Bette, and replete with a new Ford sedan - set off into the sunset for the West to begin a new life. I was almost 25, and raring to build UFO's.

ALBUQUERQUE, 1950

Sometime in the early 1800's, a very successful trapper, William Randolph Lovelace, bought 300,000 acres of barren desert land just East of the Rio Grande River for fifty cents an acre in the Territory of New Mexico. Not too much changed for the next 100 years or so. Today, however, greater Albuquerque sits on this land, and Lovelace's heirs do not have to trap for a living. When I first arrived in Albuquerque in late '49, the population was short of 39,000; when I left six years later it was about 125,000; and now, in 2009, it is over 500,000. Modern Albuquerque was born of the Atomic Age.

Near as I can tell, the city's early growth from an army fort, a trading post, and a haven for several Native American pueblos, was based on its central location in the territory/state, its fertile land straddling the Rio Grande and, as a consequence of its being a way stop on the Santa Fe trail and a railhead for the Santa Fe railroad line, a center of trade and financing. Prior to the discovery of vast uranium fields in nearby Grants, it was a bilingual state with "Anglos" in the minority and Democrats in the majority. The State Capitol, Santa Fe, -about 50 miles North of Albuquerque was a very old historic town of around 15,000.

The great beauty of the state, especially in the North above Albuquerque was generally unappreciated by all outlanders but true western aficionados. Most outsiders were not sure it was a part of the United States. In fact, even today, the *NEW MEXICO* magazine always features a humorous article entitled, "One of our Fifty is Missing" which points out the fact that many American are not aware that it is truly part of the Union. A few adventurers came for sightseeing vacations, fishing, or dude ranching. Its outstanding attractions in art, opera, architecture, museums, skiing, fishing, rafting and nature, as exemplified by White Sands and Carlsbad Caverns, were not as apparent or heralded as they are today. Nor was the use of its primordial salt beds as a site for nuclear waste dumping.

When we arrived to settle in Albuquerque, the state was bilingual, and generally run by people of Hispanic origin. Ordinances, ballots, and all State's business papers were all printed in both English and Spanish. If you didn't speak a modicum of Spanish, you could run into language difficulties outside of the city. Agricultural pursuits were the main business - alfalfa, soy beans, cotton and grazing. Native Americans made up a significant portion of the population and, though respected, were pretty much left to their own devices since they were not part of the political equation.

A man named Clinton Anderson; a realtor, newspaper owner and publisher, and power company partner had been elected US Senator. He already wielded considerable power in Washington, when he caused the Atomic Energy Commission to come into being and cajoled Congress to accelerate the progress in all things nuclear, using the "Iron Curtain" as the prod. The arrival of the atomic age changed the face of New Mexico forever.

Having supported the establishment of *Los Alamos* where the bomb was first developed, Anderson now directed his energy towards making Albuquerque the center of the burgeoning atomic energy business, particularly the weapons end of the spectrum. He caused congress to spend untold (i.e., hidden-in-the-cloak-of-secrecy) millions of dollars to build an atomic weapon arsenal at top speed. Albuquerque exploded into a boom town - perhaps the most violent such explosion of all time. The town had been concentrated around the quaint Mexican traditional downtown "Old Town" area and north along the river valley toward Bernalillo, the county seat, and east out to the University area. Now, the northeast valley area literally grew up overnight on gritty desert land in the vicinity of Sandia Base, where the bombs were being designed and weapons effects evaluated. The sounds of bulldozing and hammering overwhelmed the atmosphere day-after-day. New neighborhoods and shopping malls were being created as if by magic.

The breezy air was filled with the fine dust that Alan Pope, whom you will hear about later, so eloquently sang about in his epic, "*Watch the Land of Enchantment Go By*". You could taste the grit in your mouth and you needed a lot of water. You wanted to take two or three showers a day, because your body felt like it was covered with a fine emery paper. The road, Wyoming Avenue, from Central Avenue, Route 66, to the Sandia Base entrance kiosk was finally paved. Everybody, including the Native Americans had jobs, and the empowerment and confidence that comes with prosperity. Hallelujah ! -WE WERE MAKING THE BOMB!

WHAT! NO FLYING SAUCERS?

None of my professors or friends at NYU knew what Sandia Corporation's mission in the scheme of things was. We all knew it was secret government work, and some speculated that it might have something to do with the Los Alamos Lab, also in New Mexico. I was pretty sure that they were making flying saucers. I went there with high expectations of being in on the ground floor of the next big thing.

I was surprised, and somewhat chagrined, but not morally outraged, to find out that the true mission of Sandia was to design, build, and test atomic bombs. The warheads were designed and built by LASL, the Los Alamos Scientific Lab, at a then - sequestered community about 60 miles northwest of Albuquerque, nestled in the lovely Jemez mountains. Drop-testing of instrumented 'dummy' bombs was assigned to a joint military test force which included the 4925th Strategic Air Force Squadron along with other support elements of the Defense Department and the AEC, at Kirtland Air Force Base, located West of and adjacent to Sandia Base. Los Alamos had the lead for live bomb testing in the Pacific. Testing of inert bombs for flying quality, trajectory correctness, and fusing and firing ability was Sandia's responsibility. Drop tests were performed mostly into the Salton Sea in California, with 'arctic' tests into Lake Bemidji in Minnesota.

After getting over the initial shock, and filling out the myriad forms needed to obtain the 'Q' clearance necessary for me to do the job I was hired to do, we began the settling in process. For the first two weeks, the mile high altitude took its toll. We needed a lot of sleep to acclimatize.

My wife had to learn pressure cooker techniques, and our car's carburetor had to have high altitude jets installed to operate smoothly. We rented a three bedroom house on Adams Street, a few blocks North of Central Avenue in the near heights, and were promised that the road would soon be paved. We made friends with our neighbors quickly - again, the mid-West type friendliness that was so foreign to Easterners came into play. All the families in our neighborhood were young - in fact until our kids went to public school, I don't think they knew there was such a thing as an old person. So it is in a high tech boom town.

I started to work in the sequestered research organization's West Lab, which is at the site now occupied by the USAF's Officer's Club. The site was a former private girl's school, and it now housed Pete Petersen's Research Directorate. I was given a non-secret on-loan assignment until my 'Q' came though. I was finally off and running.

SETTLING IN AT SANDIA.

I became an engineering supervisor without being aware that I was "promoted". One day while I was busily engrossed reading a technical report about this new-fangled transonic speed range and the "Sonic Barrier" that came with it, my boss, Ken Erickson, came by and said, "Son, how'd you like to watch over some new aerodynamicists that I just hired?" Without realizing the implications, I quickly replied, "Sure, Ken, if they don't take up much of my time." With that simple naive reply, I sealed my professional fate for the next 30 years, and immediately went back to my work. Fortunately, I turned out to be a good, if not excellent, manager. At the time, however, I was very happy being a ranch hand applying the stuff I had learned in college and grad school, and probably would not have willingly changed status if I had really considered what was happening.

When I returned from the "Crossroads" Atomic Bomb drop test in the Pacific (see my earlier book, "*On the Cutting Edge*" for details), I assumed that I would continue to assist my then mentors, Drs. Mel Merritt and Jim Shreve in analyzing the data Mel and I had taken at the test. Unbeknownst to me, Ken has negotiated with Ev Cox to bring me back into his organization. He had on hand a very polyglot bunch of civilians and military personnel, and wanted to reorganize by technical discipline, thereby hopefully reducing the chaos. The military people were on stop-gap assignments from the Army and Air Force - but had A-one engineering, physics and/or math training. They were loosely under the leadership of Dr. George Hansche, a U. of Wisconsin physicist who was another member of the Johns Hopkins mafia. I had met George on my interview trip, and was impressed because he pointedly had told me that he had seen at least two UFOs which he felt were unmistakable. He would be my immediate boss for a short time.

His assignees included Army Col. "Mac" McEvoy, Lt/Col. Connie Nelson, and Col/Dr. Jim Sharp (like Ken a Texas/Austin graduate), both USAF, and Captains Wil Dondanville (Army) and Bob Pearce (Air Force). They, along with Natalie Bradley, a University of New Mexico grad with an MS Math degree, were roughly responsible for calculating the flight path of all present and under-development A-bombs. Mac was assigned by the Army to make up the operational bombing tables. These tables enabled bombardiers to signal release for all anticipated bombing run altitudes and speeds. He was a West Pointer, who eventually made General, even though he was strictly a technical officer. I think maybe they appreciated his easy way of dealing with people, while being a conscientious bear for work, and most importantly being one of the few people who knew their way around the ENIAC-like computer that the Army operated for him in Washington. Natalie assisted him in this effort- she was Mrs. Inside to his Mr. Outside. Her husband, Zorro, also a native New Mexican, was a ranger in the Tuzigoot National Forest in Arizona.

Jim Sharp acted as liaison with the Kirtland Field – based 4925th Strategic Bomb Squadron, under the command of my next-door neighbor, Col. Osmund J. (Ossie) Ritland, and the new Navy test squadron that had just become resident at Kirtland. Jim was a brusque no-nonsense Texan whose final military assignment in Albuquerque was to oversee the design and erection of a huge liquid hydrogen vacuum storage jar. This program started in 1952 as soon as it became apparent that such would be necessary in the development of the H-Bomb. After he left the service, Jim had an outstanding technical career, ending up as engineering VP of the high tech Southwest Research Lab.

The other officers were assigned to assist in determining the drag of the various bomb shapes and in assisting in aerodynamic problems. Connie Nelson and I became friendly since we shared similar educational experiences and capabilities. He had the "Engineers" degree from Cal Tech - being just short of a Doctorate. The Air Force, which had paid for his education, felt they needed him in the field before he could complete his thesis, which alas he never did. Connie was the first person I met who was hooked on "Scientology". He had his own "reader"; felt L. Ron Hubbard was the new savior; and could go back to his thoughts in the womb. Even though he knew that most of us felt that this was a great boondoggle, he was so avid that he would proselytize - given half a chance. Nevertheless, he was very likeable, with a laconic sort of wit, and he and his wife readily fell in with our bunch in our active after-work social life. Later came USAF Col. Harry Evans, destined, like Col.

Os before him, to head up the Air Force's space research and development effort at the Los Angeles Air Force Base as General Officers.

There were also a few civilians on hand, all recruited slightly before me, who also reported to George Hansche: Bob Black, a graduate of BYU, and, from UNM, Mary Jo Hazard, both mechanical engineers, formed the bulwark of the wind tunnel team, assisted by the military people. Mary Jo's then estranged husband, "Hap" was a nutty professor type innovative designer on his way to California. The next time I saw him was in the early sixties when we both worked for Space General, and he was running around the country hyping the space age, dressed in a deep sea diver-like space suit that he had designed. Then, there was Jack Lohse, an older very quiet man who was a pilot and did odd jobs in aerial instrumentation.

Ken Crowder ran a group of mechanical designers who built our wind tunnel and firing range models. Bill Pumphrey worked for him, but it was Bill's wife, Carrie, who was George's secretary, who had everyone's heart secretly aflutter. She was breathtakingly gorgeous! A milk-white skinned sultry brunette with high cheekbones, a la Gene Tierney, with one-quarter Indian blood, she was as nice and unassuming as she was beautiful. How the wives of my colleagues must have hated her! She inspired me to write my only song, which I dared not tell anyone about. It was the tear-jerking "*I'm Going to Carry Carrie from Tucumcari to Carrizozo*", which everyone will surely recognize refers to famous New Mexico hinterland towns. While driving to and from work, I sang it out loud and effectively in the authentic western nasal twang that I had learned from my boot camp buddies.

In an indirect manner, Carrie taught me the most important lesson of how to succeed in the aerospace - and probably any other endeavor - world; be nice to all secretaries, for they have the power of life and death over you. Of course, in Carrie's case, it was easy to be pleasant as many times as possible during a working day, without it being over obvious. She was easy to talk to, once you overcame your stage fright, and always asked how things were with you. After I became more comfortable with her overwhelming beauty, I kidded around with her and found my work always seemed to get done first and with high competency.

There were a few others in prior residence in addition to the notorious Hap Hazard, and after Ken Erickson assigned the two new University of Colorado master degree-in-Aero. E. graduates, Hal Vaughn and Randy Maydew, to me, I was given the green light to hire people. I was dubbed Supervisor of Theoretical Aerodynamics. My group was to be responsible for the design and flight characteristics of new bombs and for obtaining in-flight atmospheric pressure data which was utilized to explode the bombs at the desired altitude. We planned and ran the wind tunnel and firing range tests, haggling amongst ourselves about what model changes to make. At $300 per occupancy hour, this took a lot of thought and careful planning. But, it was great fun and took me to many far-away places!

After that, came the deluge. I began taking on engineers to meet the ever increasing assignments that came with trying to develop two new A-Bombs a year. I unconsciously turned into an empire builder - a characteristic that I steadfastly maintained through thick and thin through the rest of what was to be a 30 year managerial career. There came on board people that I have stayed close with for the rest of my life: the U. of Michigan mafia, Ed Clark and Warren Curry, and later-as a consultant - their classmate Arnold Ducoffe, the then Head of the Aerospace Engineering School at Georgia Tech; Bill Pepper, another Coloradoan who would rejoin my life in the 80's when he came to consult in California; Paul Rowe from Iowa State University -

who would also play a major role in my future life -, and a host of others - all present in my group before Alan Pope, who we will hear more of later, arrived. All of the above but Paul, Ed and myself were destined to stay at Sandia for the rest of their professional careers as 40-50 year men. All achieved fame in the business: Hal in rocketry and as an inventor and author in riflery and as a husband to Mary Jo; Pep, and Randy as inventors and developers of high speed parachutes; which low flying aircraft that deliver H-Bombs require to give them sufficient delay time to allow them to escape the blast. Later, Randy was sent to Spain to find the H-bomb that, as a result of a bomber mishap, had fallen into the ocean off the coast. He has written stories and a book about this adventure, as well as stories about his war experiences as a B-29 navigator. In his retired years, Pep, consulting with LA-based Irvin Airchute designed the parachute that decelerates the Space Shuttle after it touches down. These were fine, inquisitive and gung-ho people, who made work a pleasure and anxious for the next day.

I was blessed throughout most of my managerial years with wonderful secretaries (alas, George took Carrie with him when he moved out). Elvina "Viney" Strance was my first and the prototype for those to come, since I always looked for the attributes she brought to the job in all future interviews. Take competency for granted, including the ability to take dictation, which I soon learned to love and be good at. Then, you want a substitute Mother who genuinely cares for you and your flock; an iron willed disciplinarian who forthrightly tells the frivolous and low priority to go to hell with several other spicy cuss word to go with it; a sense of humor even under stress; and a collector and raconteur of far flung rumors and spicy gossip. Viney was all of these and more. She will remain in my heart forever, as will her successor, Emily Edwards, who came after Viney was promoted to be Ken's secretary. I saw Viney a few years ago. She bitched about being old - but was obviously as starchy as ever and still got a big kick out of life. Emily was a sweet, elderly, motherly type, but very efficient and quite knowledgeable about the group's doings. I kept in touch with her after she retired in Idaho until her death at a respectable age. So, life was good; life was interesting, and we were clearly winning the Cold War.

More new recruits joined the group, and even though from diverse back- grounds, all seemed to like the informality of the Southwest and the tremendous esprit de corps. We became a big happy family; wives and children included. Wife #1 and I, having rented 405 N. Adams for about a year, bought larger 410 across the street, next to Jean and Colonel Os. We needed more room, since we had added son Bobby to our family. We almost were able to make the down payment on our own, lacking only $800 dollars, which my father nicely came up with. The mortgage payment was actually a little less than our prior rent. In those days, houses were easily affordable - a far cry from the nineties when our kids really had to struggle to acquire a decent house. Albuquerque became a full-fledged boom-town; with full employment, great weather - except for the flying sand, and much joie de vivre. We liked people in general, and the ones I worked with in particular. Parties abounded. Every weekend one of our 'gang' would have a Bar-B-Q at their house or a long cocktail hour affair. Work was fun with people you liked!

New Mexico remained the plaything of Senator (D, NM) Clinton Anderson who had won oversight over the AEC, among other political plums. The Public Service Company of New Mexico - which provided power to the state from its plants in the Four Corners area - fuelled the Democratic political apparatus with as much money as required. I had been active in Democratic politics almost from the moment I moved to New York as a jazz musician/graduate student in the late 40s. On my arrival in NYC's Greenwich Village, the first visitor I had to my room was Sam Resnick, the local Tammany Hall ward heeler, who befriended me and provided me with work addressing mailings, at $6 per hour, whenever I needed some dough. I enjoyed "getting out the vote" in the neighborhood.

There was no doubt that Tammany Hall ran New York – I found that out early on. One Saturday morning, Sam invited me on a leisurely stroll to make the "rounds" of local businesses under his "sponsorship". The shop owners were always glad to see him and quite often, without asking, they would proffer a contribution to the Party. "Why do they do that, Sam?", I asked. He explained, like a 'Dutch Uncle', that if they didn't contribute on a reasonably regular basis, they might be visited by the fire inspector, the insurance adjuster, the Department of Health, and so forth - and might be subject to fines. I always felt that the fall of Tammany was a tragic milestone of a bygone age. They really knew how to run a city to assure self-preservation and spread the wealth (again, doing well while doing good)!

Quite naturally, we continued our active Democratic political proclivities in New Mexico, again in a caring and protective atmosphere. By acclamation, and with my wife's help, I became precinct Chairman of a rather densely populated six square block residential section of the town in our immediate neighborhood. We were encouraged by the Party to have monthly pre-election meetings, which were held at our house. Food and drinks and Republican-knocking abounded. We were always reimbursed for our expenses, no questions asked. Shades of Tammany and the Kansa City organization of Tom Pendergast!

I was a delegate to the state political convention held in Albuquerque to draft the Party platform for the upcoming national election in 1952. As Chairman of what surprisingly enough was the largest precinct in the state, I already had some notoriety for the good parties we held. Now, I attracted some statewide publicity for advocating a plank, which was vigorously supported, to eliminate prejudice against the many Native Americans who populated the state in Pueblos or Reservations. My reward for this notoriety was my selection to become an alternate delegate to the national convention in Chicago. This was the site of Adlai Stevenson's first run for the Presidency, and he was adored by the flock, and certainly by the Brodsky family. So, I attended the exciting and dramatic affair, and loved it! I was back in the "Toddlin' " town of my notorious Navy career; this time living in relative luxury- all expenses paid - instead of residing in a third tier bunk at the nearby Navy Pier.

The convention was exhilarating - nothing could stand in Adlai's way to the Presidency! He was erudite, witty and wise. The famous picture of the worn out holey sole of his shoe marked him as a man of the people, and was used as a symbol in his election campaign. How could he possibly lose? I returned home, unfortunately without having the time to check out my favorite war - time watering hole, the Golden Dome bar with its singing bartenders who warbled really dirty songs with great relish, but sure that Adlai would take the win. To this day, I stay in wonderment that Ike beat him! My work continued to be exciting. I was attending late afternoon classes at the University of New Mexico going for a Master's degree in Applied Math. I did this both for my love of mathematics and for not wanting to lose any of my still unused G.I. Bill of Rights school eligibility. At the same time, I was also helping out in little theater back stage work, since my wife was resuming her earlier- developed acting avocation.

I was concurrently learning a hard lesson on how to write technical memos which could be understood by non-technical people. The technique did not come easy to me, but fate, in the form of a cruel but caring boss, stepped in to help. Unlike my colleague Alan Pope, who seemed to have a new idea every day, I consistently had about two good ideas a year. One of my early ones was to propose and demonstrate a new procedure to find, in the wind tunnel, the aerodynamic forces acting on a bomb that had just been released from the bomb bay or from under the wing or fuselage of a delivery aircraft. If these forces were known, the trajectory of the bomb could be determined assuring that it would not hit the aircraft that had just released it. I wrote a detailed paper on the new technique, only to have my interim boss, Dr. George, in his usual staid and fatherly way, bounce it back for rewrite. "For heaven's sake, Bob", the very mild mannered George said, "How do you expect anyone in their right mind to understand this gibberish? Write it so a civilian can figure out what's going on!" So, I rehashed it, muttering under my breath. And, he again bounced it, through several iterations. Oh, how I hated that Man! Finally, I got it right, and could see what a tremendous improvement over earlier versions had been achieved. To this day, I thank George for being such a caring bastard. I soon began to strive for understandability in my writing, and later realized that I could deviate from the very rigid and dull engineering de rigueur writing format by adding levity and irony without losing painless disclosure. I also became more critical of my colleagues work, and hoped they appreciated my criticism as much as I did George's. But, alas, to this day, my writing is still somewhat stilted as I always prefer a high brow or multi-syllabic word to a simple one. As with many things, I'm now too old to want to change.

During the short period that George was watching over both of us, Alan Pope told me that he had also given him a really bad time with his writing. In fact, Alan said he got so mad that, having already counted up to eight in his list of ten allowable grievances against George, he said to himself, "OK, one more time and I quit!" Fortunately, right around then, Ken broke up my group into two, with Alan now being in charge of the experimentalists and me of the theoreticians. George got another assignment, and Alan and I were left on our own to slaughter the King's English as best we saw fit, which purposeful endeavor on my part continues to this very day.

BECOMING LANDED GENTRY

Albuquerque was growing by leaps and bounds, powered by the ever increasing defense efforts at Sandia and the Air Force and the readily available big money provided by the AEC. The population demographics were changing radically. The state of New Mexico was becoming more Anglo than Spanish; was suddenly no longer legally bi-lingual; and was booming. Most of the newcomers were young engineers, lured by the promise of inexpensive single dwelling houses and one month vacations. The East valley, North of Sandia Base, was being overrun by such housing projects. Adobe construction, long the mainstay in the state, was too expensive, and had given way to faux New Mexico style-wooden frames covered with ersatz adobe finishes. Still, they were very nice houses, with several bedrooms to accommodate fast growing families.

Our kids were growing up without knowing that there were such beings as people over 35, or that white was not the only possible people color. Our friends were almost exclusively confined to Sandia and military service colleagues. With the influx of new technicians came the conservatism in thinking that somehow must be ingrained in an engineer's training, unless you are from the northeast and are Jewish.

Thus, the seeds of the evil Republican party thinking and, eventually, power came to New Mexico. But there was no inflation. Our favorite meal, when we could get away from the kids, was to get a steak dinner at Baxter's, at the edge of Old Town. It cost $3! When Wife #1 inherited some money as a result of her favorite aunt's passing, we decided to invest it in a move to the north Rio Grande valley. This was the area where the local landed gentry lived, as well as some our new found friends from the Albuquerque little theater group with which we were now performing. It afforded the possibility of owning horses, which we both desired. We found a gem of a ranch house off of partially paved Guadalupe Trail, in the 7500 North section, just West of N. 4th St.

It was a large thick-walled white adobe ranch house on 10 plus acres of bottom land about 100 yards from the river levee. The walls kept it warm in the winter and cool in the summer, as true adobe is wont to do. The property periphery was completely fenced in with white wooden fences, and alfalfa was the crop. The 4 acre front field included a few fruit trees and was bounded by a long unpaved road that led to the house and the back outhouses. The unfenced area around the house included a small light farming area and parking spaces. The 6 acre back field abutted on the outbuildings which consisted of a 2-car, 1 tractor garage, stables for two horses, and two large chicken coops. We inherited about 24 laying chickens, some being Rhode Island Reds, and an ancient but working John Deere tractor with various plowing attachments. We were both city kids, and this was like a wonderland!

One of the first things we did was to buy PeeWee and Chili from our friends, the Wunderlich's, who were leaving for Sunnyvale where Bill, a mechanical designer, had been offered a job at Lockheed working on a Polaris submarine ballistic missile precursor. Bill and Margaret were moving from a small ranch in Tijeras Canyon and were seeking a good home for their beloved, albeit un-pedigreed, horses.

The horses were brothers, acquired as colts, and assiduously trained for English saddles, since the Wunderlichs, like us, were Easterners and had a similar disdain for western saddles and riding techniques. The horses were a gentle pair, very good with the children, loved to eat apples off the several trees that bordered the front field where they grazed on the alfalfa stubble during the non-growing periods. We generally nurtured two crops per summer, and during the growing seasons we fed them with our baled alfalfa, supplemented with oats. PeeWee was my horse, and bedeviled me by not responding when I called for him for an almost daily after work ride. I screamed and yelled at him and chased him with my broad-brimmed hat a'waving until he finally gave in. Wife #1 no such problem with Chili, and blamed his attitude on my "overzealous spurring". She always looked good in the saddle and always had on the proper attire: jodhpurs, helmet, crop. She could do some dressage exercises and had a history of "showing" back East. On the other hand, although trained at a very young age by my late (she died of Leukemia at age 18) sister, an excellent horse "woman", I rode hard and fearlessly, but sometimes with my back not quite straight and my elbows flapping.

The riding was great! We would mount and proceed to the back of our property. I would lean over and open, and then close, a gate to a pathway that ran alongside a tributary irrigation ditch towards the main ditch. This ran parallel and close to the river, protected by the intervening levee. There was a makeshift wooden bridge that crossed the ditch and led us on to the levee. Here we had a unencumbered medium hard dirt/clay perfect bridle path that extended 2-3 miles each way along the green banks of the mighty Rio, reduced to a sluggish, sand bar pitted flow in the summer, when all the water was in the many-faceted irrigation ditch system.

Now, in the new surroundings and school, the kids could get a much better feel for the real world of the Southwest - replete with middle aged and old folks and Spanish- and Indian- speaking friends. By prior arrangement, Joe, a true chief of a local Indian tribe, came along every afternoon around 5:30, in his battered pick up truck, to cart the garbage, tend to the foliage, and most importantly, watch over the kids, who were "assigned" to help him and learn some Indian lore in the bargain. So, we rode with easy minds. When we came back, or before Joe showed up, the kids had their turns to ride within the fenced fields. Bette and Bobby loved the horses, and our dog, BooWoo #1, and considered them all part of the family.

Where the road down our long driveway reached the house, it bordered on a fairly large area between the road, which still clung to the side fence, and the house. It was perfect spot for a then $50 plastic above-ground swimming pool, which stood about 3 feet high. It was a feature attraction at a memorable summer's eve housewarming Bar-B-Q party we gave for the group and assorted friends and bosses. It turned out to be a raucous and drunken affair. Mild "old" (he seemed so to most of us, although he was only about 10 years our senior) and quiet George Hansche had the temerity to climb up on the roof abutment that overlooked the pool, and jumped in. He lost a prized tooth crown for his derring-do, but forever gained a fame he would never have elsewise achieved. This started a poolward trend for many in various states of dress and undress.

As the evening roiled on, Ken and I decided to go for a bareback horseback ride. I let him ride PeeWee, the more substantial of the two small steeds, since Ken was a big man under his 10 gallon Stetson. We rode around the peaceful neighborhood yelling that "The British are Coming!" Then we decided to visit our across-the-Trail neighbors and Sandia colleague, Bill and Marilyn Gardner. Ken fiercely walked PeeWee up a small flight of stairs leading on to their front patio. Leaning low to avoid the front door rain shield, he banged on the door and scared the hell out of the now- awakened Gardner's. Bill had the good sense to seize up the situation, grab his antique blunderbuss, and threaten Ken with meeting his final reward. We retreated hurriedly, and the party gradually broke up, but was the talk of then town for several days afterward. The thought of it still brings a smile to my face.

PeeWee and Chili were our means to become acquainted with local society, some of whom joined us in riding while others welcomed us as they sat in their trail-accessible outside verandas at cocktail hour. Our immediate neighbors, Ginny and Bob - whose ranch was between us and the levee- rode with us and later gained fame as the developers of the Sandia Crest sky lift with its bottom and top restaurants; the latter providing access to the nearby ski lifts. Next door to us was a famous woodblocker and lithographist, Ada (Adya) Junkers. His big highs occurred when he found an ancient piece of wood, such as the one he ran off to salvage in Italy, when he heard of an old wooden bridge that was being dismantled. Other social stops included the Bob Poole's (he was second in command at Sandia) and, on the weekends, the John Simms'. John came from a politically prominent New Mexico family and headed a downtown law firm. He was very charismatic- a JFK precursor type - and we early-on got on his bandwagon with our newly found political savvy when he decided to run for Governor- a position he attained and filled with great effectiveness. These two latter stops were generally for drinks and gossip, although the Simms sometimes rode with us.

The other social outlet came as a result of my hiring myself out, at $10 an hour, as a tractor jockey. I loved driving our wreck of a tractor and, on our own fields, learned to plow a reasonably straight furrow. The big farmers in the neighborhood either refused to work on the small 10-20 acre ranches, or would charge outrageously. My services filled the bill, and I spent considerable weekend time nursing the old machine through its paces during the two annual plowing seasons. In the spring of '98, after reviewing this chapter, colleague Alan Pope wrote; "You omitted one of your greatest lines: When you bought the farm you told us, 'It also came with a tractor and a whole set of things that you drag behind it' - thus illustrating your keen understanding of farming."

We had the normal farm tragedies. The tractor needed continual nursing and, once, a new head gasket. I managed to do this repair myself with some help from Bill Gardner. Our whole Rhode Island Red chicken colony was wiped out by the dreaded *coccidiosis*. Even before that sad event, which the children took as hard as we did, we figured that our eggs probably cost over a dollar a dozen more than store-bought ones; just as our home grown alfalfa bales did. I used the farm operation as an income tax write-off, and as long as I showed the IRS an ever decreasing deficit, got away with it. In New Mexico, at that time, at least two hundred acres were needed to make it as a farmer of alfalfa or soy beans. Irrigation ditch operation was another athletic event that we cherished. Our local "ditch runner" would give us 3-4 days warning that the irrigation water would be ours for a two hour period, generally every two weeks. The appointed hours were usually in the middle of the night, and you weren't allowed - by common law - to complain about the time slot assignment. In retrospect, I think a little money may have changed hands with the runner to assure a favorable time. Our time assignments were hardly ever favorable and time swapping was too time-consuming to pursue. The alarm clock woke us up in the middle of the night and we donned our grubbies and hip boots, grabbed our flashlights and shovels, and sleepily trod to the now-unlocked weir that we opened to let the water into our ditches. We then followed the water as it flowed into our property, using the shovels to repair ditch sides flattened by horse hooves or kids. It took about an hour to irrigate our fields, whereupon we returned to the weir and turned the wheel which lowered the wooden dam. This was life in the raw!

In past years, I have revisited the old homestead. It remains as idyllic as I remember it, and the neighborhood, except for the paved side roads, remains as rurally pristine it was then. But, all things considered, an engineer's lot is probably better than a farmer's.

A PAEAN TO ALAN POPE
(WATCHING THE LAND OF ENCHANTMENT GO BY)

Those fledgling Southwest days established many life patterns and survival mechanisms for me and my family. I learned a lot of things in Albuquerque: how to - plow a field and grow alfalfa and raise chickens; - be a scientist; how to be an engineering manager; - be a husband, father, and homeowner; - enjoy and cherish co-workers; - design and test atomic bombs, and, how it was to grow up in the newly evolving nuclear age. Then, as a bonus, - gain a new friend-for-life.

One day in late '51, Ken Erickson threw an employee application on my desk, and with a funny inflection in his Texas accent said, "I think you may be interested in this". Sure enough I was! Believe it or not, it was from one of my graduate-student-days heroes, A*L*A*N* P*O*P*E*, author of the book "*Wind Tunnel Testing*" - then becoming a standard college aeronautical engineering text. I couldn't imagine why such a famous person would want to give up his tenured Professorship at prestigious Georgia Tech to come to a virtually unknown outpost. My personal philosophy had already encompassed the "gift horse" thinking. I immediately pulled out all stops to woo him West - even though I realized that my job as Division Head might be jeopardized as soon as he proved his salt. To my delight, in a rather short time, Ken told me Alan had accepted our offer - at what I feel sure was a salary in excess of mine.

I had a hard time deciding how to title this Alan Pope story. It was tempting to use some snatches from Tom Lehrer's raucous classic, "*The Wild West is Where I Want to Be*". Another possibility was "*The Albuquerque Spring Song*" (Watching the Land of Enchantment Go By). This was Alan's glowing tribute to the boom town building of housing projects in the East valley. It noted that the normally gusty spring air was laden with sandy grit. The ground was being churned up by the seemingly thousands of simultaneously operating 'dozers' engaged in extending greater Albuquerque eastward toward the mountains. I settled for it.

Alan was a delight, both to work with and as a friend (even though he was 10 years older than I and the rest of our young crew of engineers). He turned out to be a high energy, gung-ho, can-do type, who had at least one new idea a day. These turned into maybe two-to-three great ideas a year- and took a lot of my time to sort the wheat from the chaff. Then, I had to let him down gently on the wild hair ones - a task I must have done very well since he never called me a damned fool. He did find ingenious solutions to several of our major problems, and put our group into the high speed wind tunnel business right on campus. When the group grew too large, I kept the Analysts and Ken appointed Alan, with my blessing, to form a new division of test personnel. I honestly don't think we ever looked on each other as rivals, even though it was apparent that our organization, as it continued to grow by leaps and bounds, would move up in the organizational hierarchy and Ken, a physicist, would step aside in favor of one of us.

I decided that a psalm in praise of my life-long and now (2003) deceased friend, Alan Pope, was clearly in order, and might also supply additional back ground for the continuation of this saga. In addition to being a top notch engineer, Alan was a fabulous song writer and a mediocre-plus piano player. Perhaps his greatest triumph, the heart-wrenching "*Albuquerque Spring Song*", whose words are daringly reprinted at the end of this bit, was inspired by a weather onslaught shortly after he arrived at Sandia, circa 1952. But did you know that he is also the writer of those other two 'world-famous' evergreens, "*I'd Die for Old G.E.*" and "*Somebody Beeg Come Here from Bell*"? All were written under his nom de plume of 'C. Nile DeKay'. The General Electric ballad notes that you can sell your soul to the company store, viz: "What a laundry I've got in the basement, I've been paying since nineteen oh three; I can't get enough stuff for my laundry, I'd die for old G.E.". "**Somebody Beeg ......**", with its punctuated Caribbean ("Collapso", he named it) rhythm, was written for the band that played in a skit (I played cornet) we collaborated on. It celebrated the arrival of our new big boss, freshly imported from the fabled Bell Labs, just after they took over the operation of Sandia from their parent company, Western Electric. To give you some flavor of this epic, "Dere's a new job open- vary nice; You don't have to werk, just geeve advice; Rogg on de floor and de job pay well; but what good is that, there'll come a man from Bell" **This, in two poignant stanzas, expressed the feelings of the native Sandians who were constantly seeing their promotion aspirations going up in Eastern-flavored smoke.**

That did happen after I left Sandia. Alan, for over thirty years, wisely and successfully led the 80-person group which I started and in doing so garnered many awards and received much recognition from his peers. I maintain visions of him - slim and wiry, black hair and thin moustache, good looking, slight southern accent, bouncing around our offices and playing tennis with great skill (he continued to play singles late in life) He enjoyed every day in his life – right up to his passing. He taught me that engineering could make good use of excessive imagination if carefully sifted and could be great fun if you didn't take yourself too seriously.

ALAN'S FIRST WIFE, CAROLINE, DIED A FEW YEARS AFTER I LEFT SANDIA. WE JOINED THE NEWLY WEDS IN THEIR PUNTA GORDO HOME

When I got well established in the aerospace business, I adopted his credo with zest.

Shortly after we visited Alan and his new wife Ethel in Punta Gordo, Florida at the end of 1996, he sent a letter with the following note: "I think I left you with the impression that I only came to Sandia for the money. It was not that simple. I was paid $ 5000 by the school, made $ 5000 salary running the tunnel, and $ 5000 from my share in the Dutton, Pope, and Bricker, Inc. that handled the tunnel. We had three kids and lived on $700 a month with two maids!! All the tunnel-associated income disappeared at the end of the war and I had to get another job. Sandia finally offered me (at my insistence) an interview trip for free, and I fell in love with Albuquerque on the first day I was there. Nobody ever told me what I would do, but that they needed a wind tunnel man and they would pay $ 8200."

"You and Ken were easy to get along with, but Hansche wasn't. So I was tickled pink with the prospect of living in the west and working with you and Ken. When I got there I found Hansche was my boss and I kept a list of things I didn't agree with. At #10, I was leaving. When we bought special wind tunnel models for some other Department, they had no place to store them, and Hansche wouldn't let us do it (We did so anyhow). And he scared the new men by wanting to know why research was necessary and did they know the test would work. (We wouldn't have to do it if we knew.) Then, as I got to #8, you became my boss and I couldn't have been happier.

So, I HAD to get another job, and you made the West look like fun, especially with a cook-out for me and with no mosquitoes."

ATOM BOMB STORIES

ALAN'S ALBUQUERQUE SPRING SONG

Watch the Land of Enchantment Go By
(by C. Nile DeKay, Feb. 1952)

Gonna put on my goggles,
    Tape up the windows,
        Get out my bottle of Rye.
While I watch the breeze,
    I'll sit at my ease, and
        Watch the Land of Enchantment go by.

Gonna close up my chimney,
    Stop up my keyholes, and
        When the wind starts to sigh,
There's no one to sue,
    So all you can do
        Is watch the Land of Enchantment go by.

The truth of the matter its good land
    That's fertile and easy to till
My only complaint of this good land,
    Is why don't the darn stuff stay still?

When an easterly wind blows
    Nobody quite knows
        How much will blow in your eye
You'll generally find
    That only the blind
        Can't see the Land of Enchantment go by

CATCH A ROCKET PLANE

There's an Eskimo I know,
    Used to the deep snow,
        Drifts nearly twenty feet high
But now he lives here
    And just quakes with fear
        when the Land of Enchantment goes by

Now a good buddy of mine
    Thinks that it's all fine
        Praises the place to the sky
You know what I hear?
    He travels all year
        Ain't seen the Land of Enchantment go by.

I'll have to be honest and tell you
    That here you have plenty of space, but
I'd rather the ground would all stay down
    Than blow all around in my face!

So, now no use complaining.
    It's never raining;
        You get more than you buy
So stay here a year
    and conquer your fear
        When the Land of Enchantment goes by

SOMETIMES A GREAT NOTION*

The pressure at work was reaching a crescendo stage. Contrary to Alan Pope's studied pronouncement that "the Russkies won't have a usable Bomb for 10 years", the wily Soviet rascals had just announced their first A-test. We no longer

---

* from "Goodnight Irene" by folksinger Leadbelly: "Sometimes I get a great notion, to jump into the 'rivore' and drown"

had a monopoly on "massive retaliation" and had to come up with a topper to stay ahead of them. A group of young lions, led by Edward Teller, had already broken away from the conservative Los Alamos lab to develop a new type of Hydrogen bomb at the Livermore Lab in California.

My group was now in a battle against time to find a workable flight configuration for the first H-bomb, the TX-17, that did not require a parachute system to stabilize it. There was also continued pressure from the new guided missile community who were adapting atomic warheads to their Matador, Snark, and Navajo long range cruise missiles. On the near horizon, there also loomed similar tasks for the new ICBM and IRBM ballistic missiles that were being developed by the Army and the Air Force.

On one trip to California for wind tunnel testing, Alan Pope, Ed Clark, and I dropped in, at his invitation, to visit Dr. Milt Clauser, who was in charge of a small aerodynamics group in a brand new company, called Ramo-Wooldridge. I knew Milt's brother, Francis, a very well known fluid dynamicist/professor at Johns Hopkins, which explained my invitation. When we made the visit, they were located in a former church on Manchester Avenue, not far from LAX. Already on board was Dr. George Solomon, destined to become my super boss in the 80's at TRW, which is what RW became when they (now named STL - Space Technology Laboratories) merged with Thompson Products, Inc. of Cleveland, to become TRW. Later, Milt offered me a job.

My reasons for turning down his offer to get in on the ground floor of the new ballistic missile development which Ramo-Wooldridge was directing for the Air Force, were not complicated. I, like Alan, was pretty happy with the job, even though I knew that I might, being 10 years in age junior to Alan, very likely lose out to him on succeeding to the Department headship when Ken left. I was also finishing up on Master of Science degree work at the University of New Mexico, thereby using up all my WWII GI Bill education credits. Finally, with two youngsters to support, I was afraid to take a chance on a seemingly fly-by-night outfit like RW, located as it was amidst the smog and already clogged freeways; the very antithesis of the peaceful sparsely populated bucolic North Valley.

However, in the back of my mind, the lure of the new field of guided missilery sang a siren song that I knew I must eventually listen to seriously. It turns out that this was the first of three major decisions that would shape my life. As with the other two, I often wonder, "What if?.." Alan, on the other hand, had three kids in the Albuquerque school system and, I suspect, had more confidence than me that he would inherit Ken's empire. RW was never an option to him.

This trip was memorable on two other accounts: meeting people who would also have strong influences on my later life. We did much of our wind tunnel testing in the Southern California Cooperative Wind Tunnel. CWT had been built and underwritten by several southern California airplane companies, and was operated by the Aeronautical Engineering Department of Cal Tech, whose head was Dr. Clark Millikan. Clark was a tall, ruggedly handsome though slightly pock-marked, mustachioed son of Nobelist and former Cal Tech President, Robert Millikan. He definitely was not a typical academic. He drove a sports car, had a lovely and friendly high society wife ensconced in an equally lovely estate, and shared my love of hot jazz music. He was an idol of mine, since his circa 1946 book on airplane aerodynamics was the standard first text used throughout the United States. We became friends and I generally spent some time with him whenever I came out to Pasadena for tunnel tests at either CWT, or the supersonic tunnel at the Jet Propulsion Lab. His faculty member, Fred Felberg, actually ran the tunnel facility, and he assigned Dr. Keirn Zebb to manage our tests there. Fred later became chief of all the technical groups at JPL.

On the same trip, Alan had been invited to visit an old friend, Ed Crowfoot, who had made wind tunnel models for him for the Georgia Tech tunnel which had been under Alan's charge before Ken and I lured him to Sandia. We went to a loft in Culver City, where Ed introduced us to a customer from Hughes Aircraft with whom he was working. The man was Dr. Allen Puckett, another grad school hero of mine since he was the co-author of the first English language text, always referred to as "Liepmann and Puckett", on supersonic aerodynamics. The moment turned out to be an historic one. Crowfoot and Puckett were perfecting the control system that would be used on the Hughes Falcon missile, the first successful air-to-air missile, and we watched an early simulation. Allen, who shared a lifelong love of sailing with me, went on to become head of Hughes, and to appoint me as the first visiting professor to the company, in the late 70's, when I took a Faculty Improvement Leave there from Iowa State University.

So, as I would again a few years later when I had another ground-floor job offered to me at Newport Beach, I ceded my first opportunity to live by the ocean - a lifelong desire - in favor of a more conservative course. It would not be until 1980 when I finally made a permanent beachhead on the Pacific Ocean.

TURNED OFF AND GONE *

Sometime in 1955, our idyllic existence started to unravel. Sure, things at work were going fine, and the challenges there continued to be exciting, and my empire continued to grow. Ken, hating the idea of having to make a choice, said he

---

* From *"Fine and Mellow"* by **Billie Holiday**.
Love is like a faucet, it turns off and on
Yes, love is like a faucet, it turns off and on
Sometimes when you think it's on, baby,
It has **turned off and gone**

was thinking of moving on and leaving his Department to me and Alan to run. But, at home, subtle attitude changes were occurring whose ultimate danger never became apparent to me because of my almost total engrossment with my work. Only later did I understand what happened.

When I first met Wife #1 at Nick's in The Village in 1946, just out of the Navy, I was still a "wild and crazy guy" and sometime hot jazz musician. As our courtship and subsequent marriage went on, I clearly started to "settle down" (I hate to call it "maturing"), first as an ex-musician, then as a graduate student cum Instructor in Thermodynamics, and now as a father and full time engineer. True, my wife and I shared interests in the care and riding of PeeWee and Chili and the chicken coop inhabitants, in playing tennis together, operating the "ranch", including keeping the ancient tractor going as well as seeing to reaping and baling, and in working with the local "little theater" group, all with our primary concern of nurturing Bette and Bobby. But our lovemaking pace slowed down to perfunctory forays. But, in truth, I was no longer the happy-go-lucky cat that she had married. I was definitely turning stodgy!

Our entry into the world of community theater was a tacit attempt to get our lives out of a rut. The group, headquartered in the northwest corner of "Old Town" square, was an active one. My wife had some little theater type acting in college, but it was all new to me. I found work in set building and prop changing, and it was fun. The people involved were not Sandians and thus added new viewpoints to our world. They also revived our latent interest in New Mexico politics, which I had dropped after our move to the North Valley. We continued to help John Simms on his quest towards the Governorship. As usual, I took a ribbing from my colleagues at work for my Democratic proclivities. It has always seemed to me that most engineers are inherently conservative in politics, religion, and common sense. I think that it's the "if it ain't broke, don't fix it" syndrome that their education beat into them.

I had finished my requisite class work towards a Master of Science degree in Mathematics at the University of New Mexico, and was starting to write my thesis. It was about comparing results of two highly technical mathematical techniques of solving intricate structures problems. Either of the proposed methodologies could be easily handled by the new - fangled digital computers. However, I had to give an illustration by doing the calculations by hand - a very tedious undertaking - since computers were not yet readily available. I worked at the computations by fits and starts at home in the evenings. It was interesting and engrossing. But, I was in no way prepared for the bombshell that would disrupt and radically change my life.

One day, my wife told me she was having an affair, and was so wracked with guilt that she couldn't stand the status quo anymore. I was completely taken unawares though, in retrospect, but for my naivete, I shouldn't have been. We met her paramour at the little theater group and both took a liking to him. He worked as a ticket seller at a local movie house and described himself as an orphaned wanderer.

Soon, with my knowledge, he was spending days at the ranch doing odd jobs and horseback riding - and presumably more - while the kids were in school. But the concept of adultery as applied to me and mine was so far distanced in my mind as to be inconceivable. The confession was a overwhelming shock, although I was able to grasp some of the reasons for my wife's actions with some sympathy. I was not, after all, still the same white-scarfed, open cockpit person that she had been attracted to and married. We discussed the best course of action for us and the kids to ride through the storm. I would take a leave of absence from Sandia and take the kids back to my parents in Philadelphia, where I would look for a new job and finish my thesis for a second Master's degree. She would stay in Albuquerque, try to decide what she wanted to do, and sell the ranch. Apart or together, we would start a new life.

Thirty five years later in the early '90s, my wife Pat and I stood in the shoes of my parents. Our youngest son, Jeffrey, having just finished all but his MBA thesis at San Diego State U, moved back into our Hermosa Beach home for lack of finding a job. For us, this was quite traumatic. But, can you imagine *my* parent's chagrin when I returned to the long-empty old homestead with children of four and six years in tow! I'm embarrassed to think of it now, but it did represent a safe port in the shambles that my life had suddenly become. I was haunted and dogged by feelings of inadequacy and revenge. Fortunately, my old room, my former nurse, and my family were there for me.

I enrolled the kids in the wonderful remnant of the old four room schoolhouse - grades K to 3 - down the street on Wissahickon Avenue in Germantown, and settled in to complete my Masters degree thesis and to write job letters. The peace and quiet was a welcome respite as I gathered up my resources for a new foray into the real world. The kids were well loved and had plenty of attention - but the scars of suddenly losing their mother would probably last forever. In Bobby's case, stuttering developed and to this day he has to fight the tendency. Bette came out relatively untouched and both have led exemplary and successful lives. Both, as do Pat and I, got along well with their mother, who lived near their homes in Santa Fe. I keep telling wife #2 that I spent six wonderful years with wife #1; only the 7th and last year left something to be desired. On the other hand, Patti has now gutted it out with me for over 50 years and clearly deserves a medal!

After three or four months in Philadelphia, including a trip back to New Mexico to submit my thesis and sign divorce – and house selling – papers and a concurrent job interview in California. I found a job in Pomona, at the Eastern end of L.A. County. The Convair plant there developed ship - to- air anti-aircraft missiles; the first of which was called the 'Terrier'. These missiles often went out of control just at the crucial point before intercept of high altitude targets. Since I had developed a reputation for being able to tame the gyrations of unstable bombs, I must have looked like a good bet to help with their problems. They hired me as Chief of their Aerodynamics Group- a job much like the one I was leaving - but with an increase of pay, to $1225 a month, as I recall.

The divorce settlement was amicable. New Mexico is a community property state, so we split the residue. Wife #1, still reeking with guilt, gave me temporary custody of the kids, since I could provide a stable lifestyle for them. The school term was coming to a close, and I prepared for our move further West - towards the ocean which I so dearly loved. The Convair recruiting brochure depicted Pomona, at the foot of the Mt. Baldy ski run, being a mere stone's throw from Newport Beach which my to-be boss, George Burkheimer, had shown me during my interview trip. Even though this turned out to be a hiring ploy in mileage reality, I was excited about the prospect of California dreaming and sailing, and a new life for me and the kids. They knew about Disneyland and were also raring to go. The plan was for me to go first to look for a house to buy or rent. The kids, with my former boyhood nurse, Nurnie, in tow, would follow when I was ready for them.

Before I left Philadelphia, I received a letter from Hal Vaughn:

"Friday, Jan 27 (1956)

Dear Bob:

I just heard that you will be back next week probably to terminate, and since I will be at JPL (more probes) all next week, I wanted to say goodbye. We will all miss you around here for it isn't often that a person has a chance to work for someone that takes a personal interest in the individual that you have. You have contributed to the growth of all our personalities, both through constructive criticism which I appreciate and simply through absorbing some of your wonderful sense of humor and humility. I consider you more of a friend than a supervisor and it has been a pleasure and a privilege to work for you.

Also, Mary and I would like to wish you all the luck in the world in the future. Yours sincerely,
Harold Vaughn

PS. even if you decide to stay I still mean it anyhow."

I ran across this letter when I was gathering my notes (it must be obvious by now that I am a world class pack rat) to write this chapter. As it did when I first received it, it brought a tear to my eye. So ended my New Mexico sojourn. Throughout the ensuing years I have stayed in touch with Alan Pope (dec. 2004), Mary Jo Hazard (my old wind tunnel campaign buddy) and her now husband, Hal Vaughn (dec. 2003), as well as the Curry's (Warren died in 2007), Bill Pepper and his wives, George (dec. 2006) and Helen Hansche, Paul Rowe until his untimely death, Ed (now deceased) and El Clark, my two ultra-caring secretaries, Viney Strance and, until her death, Emily Edwards, and Ken and Jewell Erickson up to her death and his subsequent suicide, and fellow author Randy Maydew and his two wives. Randy's first wife, Maxine, lived for many years with constant severe back pain which could not be cured except by taking her own life. Before his death circa 2000, he wrote two books and several magazine articles. He encouraged me to catch up with his output (and I have). Emily assiduously kept in touch and we exchanged family news items. She moved to Coeur D'Alene, where she died at a good age.

IN MEMORY

Harold R. Vaughn

BORN
June 23, 1924

DIED
October 7, 2003

MEMORIAL SERVICE
2:00 P.M. Friday
October 10, 2003
First Presbyterian Church

OFFICIATING
Reverend Dr. Paul R. Debenport

PICTURE FROM 1957; MARY JO (HAZARD), A MECHANICAL ENGINEER, USED TO GO ON WIND TUNNEL EXPEDITIONS WITH US AND RETIRED WHEN SHE MARRIED HAL – A BALLISTICIAN AND AUTHOR OF BOOKS ON GUNS. HE LIKED TO GO DEER HUNTING WITH BOW AND ARROW

They were/are all wonderful friends who helped mold my character and prepared me to successfully cope with

what my future would bring. My memory is good. I can still visualize them as they were, laughing and serious and foolish, in the good times in Albuquerque.

It is true that I was lured to New Mexico in the belief that I would be working with flying saucers. I am almost convinced that the many sightings in the Southwest in the late '40s were the real thing! I believe the aliens landed, observed us, and cloned themselves into human beings who quickly populated the State. Today, they are called Republicans – for how else can you explain the change from a 100% Democratic population in 1949 to the present chaotic melange?

During my 6-plus years at Sandia, I was responsible for the aerodynamic design and pressure measurement for fusing* of the Mark 5, Mark 6, Mark 7, Mark 8 (in collaboration with Frank Knemeyer of the Navy's China Lake, CA. Base), Mark 11, Mark 13 and Mark 17 (the first Hydrogen Bomb). These developments are discussed in my book, "On the Cutting Edge". I never knew if there were Marks 9, 10, 11, 14, 15 and 16. They might have been munitions other than droppable bombs, and thus not under the purview of Sandia, which had now become the Sandia National Laboratory and is one of the government-sponsored Labs working on Fusion, the next big major change that will mark this new century. During my time there, the bomb yields (explosive power) grew from 20 Kilotons (equivalent weight of TNT needed to get the same effect) to 20 Megatons – a factor of 1000! And we were just learning how to make H-<u>Bombs</u>.

---

* *Bombs were most effectively exploded in the air above the target – altitude being a function of yield. Radar to measure height was not permitted since it could be jammed. We measured atmospheric pressure on the bomb and correlated it with altitude to assure firing where wanted.*

Chapter 3

THE SPACE AGE COMETH

- GETTING TO CALIFORNIA
- IN SPACE, AT LAST!
- AN AMERICAN IN PARIS
- SPACE (MIS) ADVENTURES IN THE 60s
- ARRIVEDERCHI, PAREE!

PROLOG

It was the mid-fifties. I noted that guided missilery was the new frontier and I wanted in. I also yearned to get near the ocean so I could get back to sail boating. So, when I got a propaganda brochure from Convair in Pomona which showed Mt. Baldy and skiing, which I also favored, three-quarters of an inch away on the map and the great Pacific Ocean a mere 2 inches away, I thought I had hit the jackpot. Not only that, but the offer of $1225 per month represented a sizable increase - considering that my entry job, six years ago, had paid $430 per month.

Convair was the Navy's prime Fleet defensive missile supplier. Their production missile, the BW-1 Terrier, worked well against low flying enemy aircraft, but lost control just before intercept at high altitudes, usually resulting in a near miss. Since I had established a reputation for being able to stabilize unruly atomic bombs, they thought I might do well with this problem. The deal was consummated and the kids and I, sans wife #1, were California bound!

GETTING TO CALIFORNIA

The five stories in this Chapter present vignettes, in roughly chronological order, derived from the Missile Age, as it began to peak in the mid-'50s, and from the very beginnings of the Space Age. For me, the action began when I came to California, divorced from both my first wife and the atom bomb business in New Mexico. In California, the lure of the coming space age coupled with my becoming a persona-non-grata in the guided missile business at Convair soon led me to Aerojet/Azusa (everything from A to Z in the USA). Here I prospered both in the Los Angeles basin and, on a one-plus year overseas assignment in France.

This is the period – 1956-1971 - covered in "The Space Age Cometh"; the action taking place in Southern California and Paris, France. The guided missile work occurs at the Convair plant in Pomona; where fleet protection missiles called Terrier's and Tartar's, and shoulder-mounted anti-aircraft missiles, called 'Red Eye's' were developed. Then the scene shifts to the Aerojet- and Space-General companies in Azusa and El Monte, California and Paris and ends, out of Industry for the first time, in the world of Academia at Iowa State University.

My marriage having dissolved, I took a leave of absence from Sandia in order to regroup. I arrived back at Wissahickon Avenue in Germantown a Philadelphia suburb, replete with two small kids and no wife. While the kids went to the old 4-room schoolhouse down the street, I completed my UNM thesis for a second master's degree (M.S. Math) in applied mathematics and concurrently looked for a job. By late Spring, I had found one in Pomona, California. We moved into a nice rental house in Claremont, a lovely college town at the foot of Mt. Baldy. It literally reeked with the enchanting smell of orange and lemon tree blossoms (ten years later, the groves were gone, replaced by smog). I soon found a live-in housekeeper to maintain the household.

I had a self-esteem-improving 3 year run as a bachelor, dating very nice ladies, one of whom became my present wife of over 50 years and counting. I picked up on my horseback riding and began training horses owned by friends from work. New wife, Patti, took riding lessons from an ex-Olympic coach, and looked quite good astride, even though she still thinks the horses knew of her inherent fear. We both started acting in the very lively Valley Community Theater; I doing second banana roles. I was always picked for dialect parts, as I could do Southern, Irish, German, Russian and Chinese accents. For example, I portrayed Gooper, Big Daddy's somewhat dense son, in *Cat on a Hot Tin Roof* and the ingenue's Irish cop father in *The Moon is Blue* and Fritz in *I am a Camera*. Patti always had starring roles.

I played tennis - usually singles - every Saturday morning and began approaching the skill level that I once had at Cornell, where I was the low man on the team's totem pole. I became very active in Pomona Valley politics, running Gov. Edmund G. Brown (Jerry's father) campaign and Bobby Kennedy's ill-fated presidential run there. (We were at the Ambassador the night of his assassination) and I was an alternate delegate at Jack Kennedy's convention coronation at the LA Sports Center.)

We started with a small sailboat that could just hold five of us and sailed in nearby Puddingstone Dam. Later, we got a trailerable 24 foot Venture, which we launched at Alamitos Bay in the Long Beach harbor, often spending overnight on the boat. We bought a house on an acre in the boonies and put in a pool and had after-theater parties. After a few beers at night, I played my horn on the back patio. The police only came once. Life was good!

CATCH A ROCKET PLANE

The Convair plant in Pomona, California had hired me in the hope that I could solve the control problems their ship-to-air defensive guided missiles were experiencing. They hoped that because of my Sandia-gained reputation of being able to tame unruly bomb peregrinations, I might be able to work some of the same magic on their missiles, which became uncontrollable just as they were about to intercept a target aircraft. It turned out that I couldn't, because it was a guidance, not a control, problem. But I did learn the missile business, and the move clearly opened new horizons - such as getting to California with prospect of finding a new wife and resuming sailing.

The missiles being developed at Pomona were shipboard deployed and protected their ships as well other nearby ships in an armada. I was Chief of the Aerodynamics Group, which consisted of about 12 engineers and 10 computress technicians, plus two secretaries. For the first time, I had a woman aero engineer working for me - a rarity at that time. We had responsibility for the aerodynamic design and control of the family of missiles as well as providing data for their fusing function As was my atomic bomb design group in New Mexico, this bunch was also very compatible. We all got along swimmingly. One of the guys took a lot of ribbing. He had just bought the first-in-Pomona strange looking 'Bug' from Volkswagen of Germany. He argued that it was very saving on gas. We thought he was crazy. It was 1956.

IN SPACE, AT LAST!

I left Convair after two-plus years, for political reasons. I played tennis regularly with the plant manager, ex-Admiral Charles Horne, who had been a cabinet member in the Truman regime. He should have been a Democrat, but obviously wasn't. It was the year of election for Governor. The contestants were Pat Brown, a Democrat, vs. Senator William Knowland, the Republican who was known as "The Senator from China" because of his avid espousal of all things Taiwanese. He was also a near fascist in my eyes. Charlie Horne wrote an article in the Management Club's newsletter urging all loyal managers to support Knowland. On the tennis court next day, I told him angrily, "Charlie, you can't use the Management Club for politics." "Bob", he said, "I can do anything I damn please. I'm President around here." That really burned me. I sent a letter to Frank Pace, a good Democrat, who was the President of General Dynamics, of which Convair was a part, telling him what was happening. I sent a copy to Charlie. He tried to can me, but my Chief Engineer protected me. But, I had no place to go. I started looking. I wanted to go into the SPACE business.

It was a down time in the aerospace industry. There were two job possibilities: An offer to become chief of the technical staff in the Space Division of old-line company Aerojet-General in Azusa, about 10 miles from my Claremont home; and a chance to become a ground floor employee of a brand new company in Newport Beach, called Aeronutronics. We normally spent our summer vacations there at a rental. I reported for an interview at Newport Beach in a motel suite in the Jamaica Inn. There, ensconced at the beach town that we so loved so well, were the 11 original employees. They pointed up the hill and said, "We're going to build a big plant there. We've got a huge Air Force contract!" Then they talked about more money than Aerojet had offered. "Oh, sure", I said to myself, "And I've got a new wife and three kids to support." I chose Aerojet in 1958, and it was to be another 22 more years before we arrived at the beach for permanent living.

In retrospect, it is hard for me to decide which prior job yielded the most fun and growth to my career as an engineer; the formative years at Sandia where everything was new and money was no object, or the beginnings of the space age at Aerojet, where the wonderful laissez-faire attitude of management gave the lead to anyone willing to take chances and dream big dreams. The funding agencies were in cahoots with their customers - allowing cost-plus-fixed-fee contracts for research and development work of high risk. We could try out ideas that were engineering stretches without fear of bankrupting the company. The government urged us to take chances, spurred on by the "Cold War' environment. The 60's were an engineering golden age, culminated by the Moon landings. We may never see another decade like that one!

At Aerojet-General, and then at it's newly formed subsidiary, Space-General, I was involved in many unique projects described herein and in *On the Cutting Edge*, my earlier book. For most of the time, I had a wonderful Boss, Vice-President Charlie Roth. He swore like a trooper at the top of his voice, but he knew how to motivate me to work my ass off, and taught me how to write winning proposals by promising, but not guaranteeing, the Moon. I think customers like a certain amount of optimism in both technology and cost estimates. I supplied both, pushing the envelope as far as I reasonably could. I won a lot of proposals! I got promoted as a result – but my basic job of writing proposals for new work remained the same.

Once I got the hang of it, proposal writing and managing became fun – and I got very good at it. In fact I spent the majority of my time at Aerojet running proposals for advanced work, no matter what my official title was. In one run, I won six competitions in a row, which got me promoted to Chief Engineer. Even then, I continued to write them. I had learned the technique from experts – Charlie Roth and one of my consultants, Bernie Mazelsky. They taught me how to promise the moon for sixpence and still not get caught in an outright untruth. In those days of Cost-Plus-Fixed-Fee contracting, that was the winning formula! We toiled for two to three weeks of 7-day-a-week proposal efforts. Then, I would take a week off and thirst for the next assignment.

Even nicer for me, the high management of Aerojet/Space-General were friendly and for the most part Democrats. Thus, they rooted for me in my political activism, and when a cushy Corporate assignment based in Paris arose, I had no lack of champions to approve the appointment to the Corporate staff.

AN AMERICAN IN PARIS

In the early '60s, my outfit, Aerojet-General of Azusa and Sacramento, itself a part of the General Tire and Rubber Company of Akron, Ohio, bought up a small hi-tech company called Space Electronics. This Glendale company, headed by James Fletcher – destined to later become the head of NASA, and Frank Lehan – whose connections to the spook world would later draw several of my friends towards the Langley, Virginia headquarters of the CIA, was expert in space communications and allied specialties. The idea was to merge it with my Aerojet 'Space Division' outfit to form a company that could compete with the major aerospace systems companies, like Lockheed, Boeing, Martin-Marietta, and McDonnell/Douglas. In a grand splurge, Aerojet formed a subsidiary "Space-General", and built it a new campus in El Monte, across the street from the equally new 'Black Box' that would become Aerojet's Corporate office building. After running and winning several successful proposals for new business for the new conglomeration, I was promoted to Chief Engineer.

The Space-General Corporation simply did not take off and grow as had been anticipated at its inception in 1961. This, in spite of a corps of top flight engineers, we couldn't seem to win the 'big ones'. I suspect we tried to bite off more than we could chew. In 1968, after a series of management changes, it was decided to re-absorb it back into Aerojet-General, the mother company. About that time, I got a letter from my friend Howard, who was running the Aerojet corporate office in Paris, saying that after almost 5 years in Europe, he and his wife were ready to come home. He said he would back my petition to take his place, if I wanted to try for it. Buffeted about by uncertainty on the home front, desiring a new totally different experience, and after getting concurrence from my wife and 4 kids, I did!

The big adventure began on an Air France night flight from LA to Paris. I was flying First Class-for the first time in my life - because company policy dictated that if you flew Red Eye, you flew First Class. Shortly after reaching cruise altitude, La Hotesse presented me with the menu. I quickly saw that I would have no trouble in reading French despite the fact that my three years of language courses in high school were not big on food items. However, I did see that I would now face the first big decision on my new assignment. Three equally wondrous entrees were carefully described on the menu, and I pondered mightily while waiting for the stylishly dressed young lady to ask my choice. She never did. As the night went on, each entrée was served deliciously and impeccably in the order listed. Thus, I got my first true insight into l'essence Francaise. This just added to the excitement, trepidation and anticipation of what lay in store for me.

It was mid-May 1969. I was on my way towards a one month indoctrination visit to learn the ropes of my new job. I was in my eleventh year of employment at Aerojet- General / Azusa and its subsidiary, Space-General / El Monte. Recently, massive reorganizations had begun as my newly defunct outfit was being absorbed back into the parent company. With my new assignment came a transfer into the Aerojet Corporate staff. A few months earlier, seeing that the end was near, I had applied for, and won, the plush assignment of running Aerojet's European office located at 164 Avenue de Neuilly in Neuilly-sur-Seine, an integrated superb of Paris, on the northern edge of the Bois de Boulogne. The Avenue was the western extension of the Champs d'Elysees, and is now called the Ave. Charles de Gaulle. The office, and our prospective house, were both located near the Pont de Neuilly, at the Ave. de Madrid terminus of the Metro #1 line. The plan was for me to get acquainted with my new surroundings, and make the necessary arrangements for housing and schooling for my wife and four children. I would return to California in about a month, after school was out, and bring the family back to France for the duration of our tour. We had agreed that we would be willing to stay at least three years.

After my selection, my new boss, Bill Gore, the corporate International VP, told me that I had won the hotly contested Paris derby because of several factors: I was solidly backed by the four plus year job incumbent, my old friend Howard Robbins; other assets they considered were that I was even tempered, not rash; easy to talk to if not outright voluble; and a good manager who could not be bullied. They also felt that I had a certain amount of sophistication to go with my fearless smattering of French. Also, I would be addressed as Le Docteur Brodsky or Herr Doktor Brodsky - and they believed that Europeans were more impressed by titles than the average American.

More specific, however, was the fact that I was intimately involved in rocket vehicle development and had participated in the development of an unique automated medicine device which we had just developed. Management felt that the latter might be a big seller in Western Europe, and that I could build on the start that Robbins had made in the rocket technology area. However these attributes notwithstanding, I've always believed that an additional unstated factor in selecting me was that they were about to reorganize me out of my Chief Engineer's job, since there already was an incumbent in Azusa. I accepted the job readily, after getting the wholehearted endorsement of my family. The challenge was impressive: I had never been to Europe or dealt with people on an international level. I was both scared of and eager for a real change in life style.

The Paris office, as I was about to take it over, consisted of the Director, an office manager (in this case, one Gudrun Rosner) who was not only the real brains of the gang but also was fluent in English, French, and German), a consultant - in this case the defrocked former Director -, and a large highly-detailed three dimensional map of Paris glued to a wall. This small enclave had been operating gloriously in the black for several years, being credited with a few million dollars per year in pure profit from the royalties obtained from NATO's use of an Aerojet rocket motor patent and technical support covering the manufacture of the rocket motors which powered the Hawk surface-to-air missile; then the bulwark of the West's defense system.

In fact, these royalties were the prime reason that Aerojet maintained a European base, which had been established shortly after rocket motor production in Europe had begun. The other income from the technical support of French and ESRO (European Space Research Organization) experimental rocket programs, which Robbins had promoted and was leaving for me to continue, paid for our daily baguettes and office expenses. Although other ventures were in the offing, none appeared capable of approaching the NATO financial harvest anticipated from future Hawk missile royalties. The Hawk rocket motors were being manufactured under our license by the French (by the Ministry de

Poudres, Toulouse), by the Germans (at Waldkraiburg, east of Munich), and by the Italians (at Colleferro, near Rome). When I arrived, the initially contracted production run was almost over and negotiations for follow-on production were already under way.

My instructions from the International VP were few and, so I thought, relatively straightforward: Keep Aerojet's flag waving in Europe by becoming a member of the international community in Paris; take good care of all visiting firemen, be they Aerojetters, customers, or potential customers; and manage the English-to-French translation of the operation and maintenance manual documents of the satellite communication ground station we were building in Morocco; but do not ask questions about and forward, without opening, monthly statements concerning a Moroccan bank account maintained at our office's bank; support the NATO program as first priority, especially by trying to move the negotiations along; support the French and ESRO rocket programs as second priority; visit and placate M. Samuel at Compagnie Sedam once a month and, if supplied from the US, show him new pages of a revised royalty agreement between Aerojet and Sedam; try to sell our automated syphilis test machine and required serum to all Western nations; assist the French high tech development company, Bertin & Cie, on mutually undertaken proposals; and –beatifically - enjoy the experience, remembering that money was no object.

Only the 'Bank' advice puzzled me. Much later, I learned that the envelope contained 'baksheesh', a helpful monetary addition to the Moroccan government official responsible for getting power and water to the remote site, outside of Rabat where we were building the communications satellite ground station destined to free all of North Africa from the very high tariff the French telephone company that 'owned' the underwater cable between Morocco and France charged for the service. Now, long distance calls to

Europe and the rest of the world were affordable to 'plain' folks.

The next morning, Howard and Gudrun met me at Orly, whence my training began. Happily, the semi-rigorous agenda allowed for sight seeing in Paris and included attending the famous Paris Air Show at Le Bourget, where Lindy had landed many years ago. Here, I was impressed by the amount of commerce going on in the various chalets where the great aerospace companies of the world presided over vast quantities of free food and drink and, I suspect, women.

During my month introductory visit, I made arrangements to buy the Robbins' Peugeot 404 (quatre cent quatre) with its Florida license plates, and to take over the company-paid lease of their house on Cinq Ter Rue du General Henrion Bertier; learned to pronounce Neuilly almost accurately, but struggled mightily and, to this day to no avail, to correctly pronounce Roule, as in Ave de - and Hotel de- where I stayed during this initial visit. The damned silent "e" at the end somehow gets pronounced in a way that only Parisians can do. For example, I could always "hear" Gudrun's "e"s, even though she was a Bavarian by birth, but clearly now a Parisian.

I quickly learned my way around the city, aided no end by the graphic map which covered one wall of the office foyer. It turned out that for the most part, our main business could be done by walking to the nearby NATO and ESRO headquarters. I soon learned the Metro system almost by heart. By happenstance, an ego-boosting linguistic triumph occurred early in the game. I had descended into the Pont de Neuilly metro station on my way to the Louvre and almost ran over a Chinese kid. He was obviously lost, and even had his Metro map upside down. With my newly gained confidence, I went up to him and said, "Puis je vous aidez, Monsieur?" He stammered back, "I don't speak French. Do you know English?" "Certainement, Monsieur", I replied. "Where do you wish to go?" We then discussed his train options until he knew what he had to do. As we parted, he thanked me profusely and told me how well I spoke English. "Merci millefois, Monseiur. C'etait mon plaisir", I signed off without batting an eyelash or a smile.

During the month visit, Howard took me on the business "rounds". I met the NATO major domos in Paris, just before NATO Headquarters moved to Brussels. I gathered that the follow-on contract negotiations were going slowly- not so much because Aerojet had asked for a few royalty dollars more per rocket motor- but because the countries involved could not get together. They were arguing among themselves about schedules and urgency, numbers, deployment sites and technical improvements. It did not occur to me then that when the production lines stopped, so did the royalties and, consequently, the office black ink record. But I did get a kick out of socializing with and being treated deferentially by the NATO General Officers. I was not that far away from being of Seaman First Class rank in the US Navy. In my first encounter with them, I noted that the language of NATO was English, so my technical French was not yet put to the test.

We next visited the licensed rocket motor production sites. The most interesting of these was the German installation, whose factory layout had been designed by the Aerojet Solid Rocket plant in Sacramento. They had also supplied the huge batch mixers where the very unstable propellant was prepared for pouring into the motor casings. We drove there in a rental car that we had picked up at the Munich airport. Waldkraiburg is a sylvan town in the midst of a large forest, through which meanders a narrow gauge railroad. The Inn River is nearby, as is the Austrian border, and Innsbruck. We stayed at a German motel (motels were new in Germany) and paid about $2 US each for a room. We noted that an additional motel building was in the first stages of construction. A year later, when I made the visit to commemorate the end of the initial rocket motor production contract, the motel proprietor apologized profusely for charging me $2.50 per night. He blamed it on the higher than expected cost of building the expansion. I assured him my company could probably afford that added burden.

The completely camouflaged rocket motor fabrication plant itself was right out of James Bond! Entry was gained by a passage that closely paralleled the hidden ingress tunnel of the narrow gauge railway siding. The entire huge plant was covered by six feet of dirt on which grass and trees were growing. We were told that the original plant was the biggest producer of gunpowder in Germany during World War II, and the Allies never knew it existed! The sod over the roof also served the double duty of containing the effects of a possible batch mixer explosion, a not unusual occurrence at such plants. The inside of the plant was pristine, and never had a problem during my watch. Most of the officials we dealt with spoke English fluently, so I was thwarted, again, in trying out my prowess in German, learned in childhood from my German nurse and in graduate school (along with refreshing my French) in order to pass my PhD language requirements.

Such was not true in Colleferro. Only one person there at the plant spoke English, and naturally he became our guide and interpreter. I didn't dare try out my newly learned Italian, especially since, when I had tried ordering in Italian at restaurants the prior evening and at breakfast, the waiters inevitably pleaded "Please speak English!" Years later, I returned to Colleferro on an adjunct inspection trip from an International Astronautics Federation meeting taking place in Rome. Now, everyone spoke English and the greatly enlarged complex had become the leading center for Italian rocket research.

Another introductory trip we took was to the ESRO (now, ESA - European Space Agency) research and development labs in Noordwijk, Holland. This lovely North Sea beach city is about 40 miles southwest of Amsterdam. We stayed in an early version of a B&B – an old white Victorian former mansion - right at the beach. Robbins had been working with the ESRO sounding rocket group who were adapting the Aerojet scientific instrument space-experiment pointing control system, which Howard had developed, to the British "Skylark" vehicle. This sounding rocket was scheduled to carry scientific instruments into space from a launch in Woomera, Australia.

The group's supervisor was a Dutchman, Koos Leertouwer. In his group were two Englishmen, a Spaniard, and several Frenchmen, Germans, Dutchmen and Italians. As we shmoozed with him, his multi-lingual troops would knock on his door and speak with him, each in their native language. He replied in kind without losing a beat. Oh, how I envied him! I was still struggling to speak English precisely and slowly enough so that foreigners could understand my native Philadelphia tongue, let alone me trying to speak theirs.

The final first-round trip was to Bretigny-sur-Seine where CNES, the French counterpart of NASA, maintained its sounding rocket base. We drove the 404 about 30 miles south of Paris, past the then small town of Evry. Today, Evry is heavily developed with modern buildings, and is a major ESA center which controls the European launch vehicle, Ariane, and the ESA launch base near Devil's Island in French Guiana in South America. Entering Bretigny, we drove through the exquisitely drab, but highly picturesque, narrow-streeted sidewalkless medieval town, down to the ramshackle buildings that abutted what was undeniably a World War I aerodrome.

Here, Jean-Max de La Mar held court with his group devoted to carrying French-developed scientific instruments to a 5 minute or more ride in near space. Robbins was working with them adapting three of his space instrument - pointing control systems, the same system that the Noordwyjk bunch were adapting for the Skylark, into their Veronique sounding rocket. This remarkable vehicle did not require a launch tower to make sure it stayed on course as it accelerated off of the launch pad. Since the rocket's speed was too low to make them effective at the beginning of launch, the vehicle's four tail fins could not provide proper guidance at low speed. In other systems, this is overcome by the use of a launch tower. Ingeniously, the French connected wires, from the aft tip of each of the fins, which unreeled as the rocket rose. The tension in the equal length unreeling wires kept the vehicle going straight up! When, after a couple hundred of feet unreeled, Veronique was going fast enough to allow its tail fins to provide guidance, the wires disconnected and fell back to earth. Our Aerojet sounding rocket, the famous Aerobee, required a massive 150 foot tower for launch, as well as a separate powerful first stage booster rocket motor which accelerated it to the same speed, in its 150 foot constrained run up the tower, as the Veronique attained at 200-plus feet when its wires dropped away. Although the French system seemed a little like the Toonerville Trolley, it did save the cost of a tower and the cost and handling of a separate boost motor. Vive La France!

Howard told me a story of earlier watching the first Veronique flight which carried his control system as it was launched in the Algerian desert. This trip occurred before the "troubles" arose between the French and the Algerians, who sought to reclaim their country. CNES then used a launch base at Hamaguir, a stark desert test base area about 400 miles south-southwest of Algiers by the foothills of the Atlas mountain range. Howard, Jean Max, and another CNES rocketeer (Jean-Pierre Morin, who later became the first station head of the now huge ESA launch base in Guiana) landed in Algiers and were picked up by a driver from the launch base. They were shocked to see that their vehicle was a *Deux Chevaux*, a very basic bottom-of-the-line Citroen sedan whose nick-name "Two Horses" adequately describes its engine's ability. This wonderful vehicle, then selling brand new for about $600 American, had seats of cloth stretched over metal bars- much like a cot. Obviously, they were not built for comfort, but to go 400 miles through a stifling desert was the stuff heroes are made of!

Howard could not, for the life of him, see how the driver knew where he was going since whatever road they were supposed to be on was long ago covered with drifting sand. When asked, the driver said he navigated by looking for strategically placed reflectors, nailed to the tops of sticks, which were not covered by sand. During preparation for the launch, another unusual incident occurred. Howard and his bunch were proceeding to the launch pad, when another Deux Chevaux, returning from the pad, approached them. What ensued, in that vast no-man's land of a featureless desert, was the old narrow sidewalk game of trying to pass an approaching pedestrian. Both cars bobbed, feinted, and weaved, but inevitably hit head on. No one was seriously hurt, but they were stranded in the desert for a couple of hours, since these were the days before cell phones. After the ride and the crash, Howard reported the ensuing launch was anti-climactic.

The introductory phase being completed, I returned home to California to make my report to the family and get ready for our migration. As I flew back, I mused about my feelings about my first contact with a new and somewhat strange world. All during the month, I had had an almost Alice-in-Wonderland aura about seeing so many strange sights and talking to so many different types of people. In retrospect, I decided that I liked the insouciance of the French, despite the Parisians disdain of poorly spoken French and foreign mode of dress. I greatly appreciated the fact that Europe appeared to be very civilized and the people very tolerant. I thought that I was going to relish the experience ahead of me, and hoped my family would feel likewise. Now, the adventure would soon begin for all of us. My business card read, in part:

Dr. Robert F. Brodsky
Manager, European Operations
Aerojet-General Corporation

SPACE (MIS) ADVENTURES IN THE 60s

My first book of space age beginnings and feats of derring-do, "*On the Cutting Edge*", needed some hyping. I took two steps to publicize it. I turned it into a "Distinguished Lecture' under the aegis of my technical society, the AIAA (American Institute of Aeronautics and Astronautics), which I periodically present around the country. Then, spurred on by an apparent renewed interest in some of the wondrous 60s developments, I presented at the "Space 2008" conference sponsored by the AIAA in San Diego. Here is how the paper started:

" 'What goes around, comes around', And so it has come to pass that many of the far-out programs that were undertaken at the dawn of the space age are again - 50 years later – being looked at or acted upon. These include nuclear applications for propulsion and power, booster *recovery and rescue, robotic planetary exploration and sea launch. The purpose of this paper is to review a few of these so that their roots will not be forgotten."*

The paper covered some of the unusual projects that I had been involved in during the 1960's that were covered in "Cutting Edge". They included: Bidding on the Saturn S-2 Stage; Building the OV-3, an early Air Force satellite; The two inflatable reentry paragliders, IMP and FIRST; the Surveyor 'Moon Walker' and its derivative - the 'Walking Wheelchair'; an early experiment in sea launch of the Aerobee sounding rocket; and on-time delivery problems of the first DSP (early warning of ICBM firing satellite) sensor system. But perhaps the most advanced early (1961) project was the 'Moon Suit', developed shortly after President Kennedy's call for Moon landings. Although it got wonderful publicity, it was never seriously considered by NASA since it was NIH ('Not Invented Here').

But, there were other programs - now mostly forgotten – that were triumphs of audacity and engineering; but, alas, were too early for their time. These are some of the programs that are now being revived with an eye towards future Lunar and Mars journeys and settlements. The advent of exploitation of our new prowess in space flight opened up both the imaginative minds of the pioneers as well as the coffers of a government embarking on cold war one-upmanship with the USSR. No company took better advantage of this environment than Aerojet – General and its newly-founded subsidiary, Space – General (SGC). First in my capacity of Head of the Technical Staff of the Aerojet Space Division in Azusa and then as Chief Engineer of SGC in El Monte (1959-1971), got a first hand immersion in the beginnings of space.

We dabbled into everything that looked promising: The Air-TurboRocket, an air breathing engine that turned into a rocket when it left the atmosphere; Reverse Osmosis which turned sea water into potable water; The Nerva nuclear rocket engine and the Snap 8 space-borne nuclear power supply; Real-time biological agent 'sniffers'; Instrumentation to be left on the Moon; plus several failures. I called them 'False Starts in Space'

Yes, the 60s era, although replete with technology triumphs, sometimes bombed. I and/or my Aerojet cohorts were involved with some of these debacles that never quite took off. I was moved to report on some of these ill-fated programs for my technical society's "*Space 2005*" annual conference in Long Beach, CA. Here is what my introductory viewgraph looked like:

A SHORT HISTORY OF FALSE STARTS IN SPACE

BY

Prof.(ret) RF Brodsky, FAIAA
USC Astronautics and Space Science Division
TRW(ret.), Director of Technology Planning

rfoxbro@aol.com

In the talk, I reviewed seven programs, five of which– almost 50 years later – are again being revived and are briefly described below:

1. AIR-TURBO-ROCKET

This program actually predated 1960 and was initiated by Bill House. The idea was to develop an air-breathinhg gas turbine engine that could convert into a liquid rocket engine when the altitude attained precluded ingesting enough atmospheric air to maintain combustion. Then, the air intake would close and oxygen, from an on-board liquid oxygen tank, would be added to the same fuel as the gas turbine engine had utilized to become a rocket engine. A 'boiler-plate' (no attempt was made to reduce weight for the demo model) version was built and successfully tested in realistic environmental conditions. We had built a better, very advanced, mousetrap and nobody came to our door! Perhaps it was the weight of the demo that scared them away. More likely, it was the lack of a suitable application.

Recent attempts have been made to achieve 'single-stage-to-orbit' designs. Propulsion for these efforts has generally utilized separate air breathing and rocket engines. It has dawned on people that the air-turbo-rocket, done with modern technology, might be the right answer, and investigations are being undertaken.

2. NUCLEAR-POWERED DEVICES

The then new nuclear technology opened the door for developments that took advantage of the readily available and very long duration heat source that atomic energy could provide. This led to three programs that Aerojet participated in – the first and third as a prime contractor- that were all too far ahead of their time; but again are now being reviewed as necessities for where we think our space program is going:

a) *NERVA* – Aerojet teamed with the Los Alamos Scientific Lab in the development of a 75,000 pound thrust rocket engine. A nuclear core heated Hydrogen, stored as liquid Hydrogen in a separate tank, which then passed through a nozzle. The massive, heavy engine and components were built and assembled at the Jackass Flats, Nevada, test grounds. Several firings tests were made and the program judged a complete success. The next step was to flight test the engine:

b) *RIFT* – We answered a Request for Proposal (RFP) for a program called *RIFT* (Reactor in Flight Test). A NERVA-based engine and tankage were to be packaged as a rocket, taken to fairly high altitude and dropped – after a needed 'warm up' while still on board – and fired into a sub-orbital trajectory. Having developed and ground tested the engine, we thought we had a big leg up on winning the contract and thus turned in a really imposing proposal. During the proposal evaluation period, the Government decided that it would not be prudent to operate a nuclear power plant in the earth's atmosphere. For safety reasons, the program was cancelled. Also, about the same time, came the realization that there was then absolutely no known application for a 75,000 pound thrust rocket engine. This put *NERVA* in mothballs, only to be looked at again recently for application as the main thruster to get to Mars.

c) *SNAP-8* - The third program we were involved in was a nuclear-powered electricity-generating turbine-driven power supply, *SNAP-8* (as I recall, 'Space Nuclear Auxiliary Power" supply). Earlier SNAP units were isotope driven and merely supplied Watts. *SNAP-8* was designed for 75 Kilowatts – a huge amount of power in space at the time. Design required much ingenuity – in particular designing around the harsh environment produced by a nuclear reactor. The solution found required three working fluids – necessary to isolate the gas turbine from 'hot' liquids and

PHOTO - COURTESY OF AEROJET-GENERAL CORP.
THE NERVA NUCLEAR ROCKET PROGRAM WAS A JOINT AEROJET-LOS ALAMOS LAB EFFORT

gases. A big problem was in finding the correct turbine shaft bearing lubricant that would hold up over long periods in a nuclear environment. The problem was solved, a ground test model built and successfully tested. At the completion of the tests, that old bug-a-boo arose: What's the application for 75 Kw. in space? There being no reasonable answer at the time, another great engineering feat hit the dust! But, obviously, such power plants are again being looked for Moon and Mars colony support now.

3. *SURVEYOR* 'MOONWALKER'

In the early 60s, the Jet Propulsion Lab (JPL), just North of Pasadena, California, was charged with finding out what, if other than green cheese, the Moon's surface was composed of – this in order to help the design of the 'lander' that would first carry crew to the Moon. The precursor *RANGER* program took and relayed back to Earth pictures up to the moment of hard impact with the Moon's surface. Unfortunately, the pictorial results couldn't definitively define the surface composition. It appeared that it might be sandy, at least on top. But what lay underneath?

Thus evolved the JPL-led *SURVEYOR* program. This was to be a 'soft lander' equipped with scientific instrumentation to permit an analysis of the 'soil' at the landing site. In addition, the lander would also carry another set of similar instrumentation mounted on a movable, by command and direction from the Earth, vehicle. In 1962, JPL issued an RFP for the development of the movable vehicle, and we, at Aerojet's new subsidiary, Space-General, decided to take it on. We considered three means of propelling the vehicle on a surface whose composition no one knew: a wheeled vehicle; a walking vehicle; and a clam-shell-like 'flopping' vehicle which snapped open and closed to move it along. The latter was attractive because it presented a large area to the soil, and thus would not sink in, as we feared a wheeled vehicle might. In the end, we

THE SPACE AGE COMETH

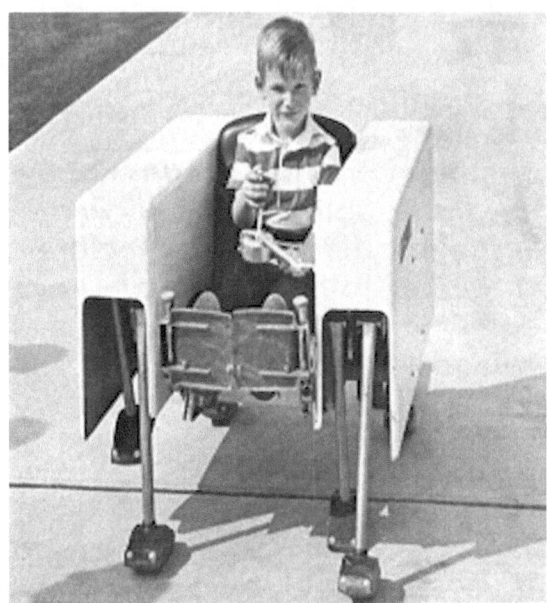

ALTHOUGH THE SURVEYOR 'MOONWALKER' PROGRAM WAS CANCELLED, IT METAMORPHOSED INTO THE 'WALKING WHEELCHAIR, AND WAS A GREAT SUCCESS.

decided that a 'walker' whose 'feet' had a sizeable area and which was capable of extracting itself (i.e. climbing out of) from a morass should surface 'sand' extend down a ways. Thus was born the Moon Walker of the illustration, and described in detail in "*On the Cutting Edge*". It was the creation of two very superior design engineers, Jack Miller and Al Morrison, both recently 'rescued' from government-level pay at JPL. We built a prototype, and Jack proudly – radio controller in hand – walked it around our campus in El Monte at noon. Everybody loved it – including the press. But then – tragedy! JPL turned us off, saying that the Surveyor had gotten so 'heavy' that it could no longer carry a separate vehicle. All was not lost, however. Ever resourceful Jack found that the UCLA prosthetics Lab was seeking a walking wheelchair, which could go up and down steps and curbs with ease. Of course, he and Al turned the -6 legged Moon Walker into a 8-legged wheelchair, three of which were turned over to the Rancho Los Amigos children's hospital and rehab center, where they lived a happy and useful life – shortened by the enactment of the law requiring all curbs at street intersections have ramps. Incidentally, the decision to use a 'walker' rather than a wheeled vehicle might not have been all that bad. In the late summer and fall of 2009, the wheeled rover "Spirit" remained bogged down in Martian 'sand' and faced abandonment.

4. SPACE LIFEBOAT

In 1962, the Dayton, Ohio, Wright Field-based USAF Materials Lab issued a very broad RFP seeking advanced programs which would 'push' technology for advanced future space projects. We were just completing a program for NASA in which we designed, built and tested an Earth reentry – at speeds much below orbital velocity – paraglider, the IMP (Inflatable Micrometeoroid Paraglider). Subsequent tests with it determined that micrometeoroids presented no real threat to manned space flight. At the same time, the German space pioneer, Von Braun, now working in Huntsville, Alabama, assured the world that a multi-crewed space station was clearly feasible by 1970. It occurred to me that such a station was akin to a cruise ship in the middle of an ocean: it would need lifeboats to rescue the crew in case of a calamity.

The design of such a rescue vehicle would certainly require new material technology; just the challenge the Air Force was looking for. Thus was born the inflatable paraglider space lifeboat, as is reported on in detail –along with IMP – in "On the Cutting Edge" as well as in *TIME* magazine in their Feb. 1, 1963 issue. Our proposal to test full-sized elements of the paraglider under realistic Earth reentry conditions was a winner and we embarked on a 1.8 million dollar program, later extended to 2.1 millions. To withstand the 2000 degree reentry temperatures as well as the internal pressure of inflation required the development of a woven thin metal fabric, two plies of which had to be spot welded over forms, and then impregnated with a thin coating of a flexible plastic material. This Dow-Corning formulation served as an ablative heat protector. It burned away during reentry, but was designed thick enough so that the metal fabric was never exposed. The program required about 5 years, and by the time we had successfully completed it, it was obvious that Von Braun's space station was not going to happen soon – NASA now starting the Space Shuttle development. Thus, again we folded our tents ahead of our time. In the 90s, with the International Space Station in business with its crew limited to 3 because that was all the passengers the Russian 'Vostok' rescue vehicle could handle, NASA again became interested in a CRV (Crew Rescue Vehicle). I tried mightily, via lectures, technical papers, and email lobbying, to get them interested in my concept. Alas, it was a victim of NIH (Not Invented Here).

Here's an example- a paper I gave in Reno in 2005:

've# SPACE STATION ESCAPE VEHICLE – 40 Years later

R.F. Brodsky,
Professor of Astronautics; Astronautics and
Space Technology Division; Fellow, AIAA
Viterbi School of Engineering,
University of Southern California
Los Angeles, California 90089-1191, USA

    Over forty years ago, the author and his associates at Space-General Corporation, a subsidiary of Aerojet-General, proposed and won what became a 2.1 million dollar contract to develop and test elements of a one person space lifeboat for crew rescue from orbit. The customer was the USAF Materials Lab. The vehicle proposed was a 1000 pound inflatable reentry paraglider utilizing a variant of the Rogallo wing design. During the life of this contract, similar one-time-usage designs, meant strictly for emergency use by up to six crew persons, were also proposed, but were not funded. This paper will disclose the work done in the 60's and will describe the author's unsuccessful, to date, campaign, starting in the early '90's, to have NASA consider its usage applied to the International Space Station (ISS).

I. **Introduction**

Shortly after the Space Age began in 1957, Wernher Von Braun predicated the presence of a manned space station in earth orbit by the early 70's. His design (Fig. 1- artist's rendition) looked like a doughnut with a central hub connected to the rim by three cylindrical spokes. The entire station rotated about the hub so that inhabitants in the rim area would feel normal gravity. Such was the enthusiasm for all things spacial, and such was Von Braun's tremendous reputation as a seer, that no one doubted that the scenario predicated by the great man would happen on schedule.

Etc.

5. PARAGLIDER APPLICATIONS

There were two other interesting applications of high speed reentry paraglider technology that, despite their promise, also bit the dust in the mid-60s:

a) MOL FILM RETURN

The MOL (Manned Orbiting Laboratory) was to be the Air Force's introductory program in manned space. Its purpose was to permit a crew of two in low Earth orbit to take pictures of the scene below. The problem, in those early –pre-digital-technology – days was that the exposed film had to be delivered back to the surface. The Air Force plan was to 'catch' the return packages, contained in a shell that could survive reentry heating, in large nets carried by a fleet of aircraft. The method was tested and it worked most of the time. We proposed that the MOL utilize reentry paragliders, which they could steer to a soft landing, to bring back the reels. At about the time that USAF started thinking about this innovation, it was decided that the Air Force would not pursue manned space. Sic transit Gloria!

b) BOOSTER RECOVERY

An even more practical application of paraglider technology was their adaptation to first -(and possibly second) stage launch vehicle booster recovery. It was standard NASA practice to recover first stage booster by parachute and subsequent ocean recovery operations. The latter action is very manpower and expensive equipment heavy. NASA Marshall Space Flight Center in Alabama put out an RFP to investigate the use of paragliders to bring the boosters back to the launch range air strip with a soft landing so that reuse would require little refurbishment. We proposed and won study contracts in which we looked at an inflatable version and a more conventional 'over head' rig, as seen in the illustration below. We proposed 'catching' the paraglider-supported booster in a large balloon-supported net which it would fly into and become enmeshed in. I'll let you guess what happened to the program? But, in 2011, I heard an Air Force Officer speaking at a Dinner meeting say that they were now looking at possibility of returning at least 1st stage boosters back to the launch area.

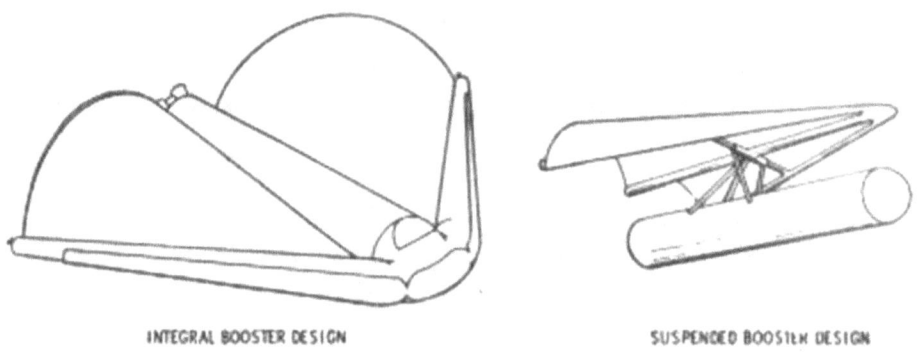

INTEGRAL BOOSTER DESIGN   SUSPENDED BOOSTER DESIGN

TWO TYPES WERE STUDIED; INFLATABLE (LEFT) AND SEMI-RIGID. BOTH WERE 'CAUGHT' IN A LARGE, BALLOON-SUPPORTED NET LOCATED AT THE LAUNCH SITE

But, I think you'll agree that all these efforts were very technologically challenging and all had a shot at or did work. All were illustrative of the many ingenious projects that arose in the 'wild west' 60s – a great time to be an engineer in the aerospace field!

ARRIVEDERCHI, PAREE!

Aerojet had sent me to France to assist our French and NATO friends adapt our sounding rocket scientific instrument pointing system to their experimental rockets. During my time there, I observed that the Europeans would buy a few systems of high tech gear from us for a high price; study them, and then make their own versions. In our case they did this very well, producing a control system twice as big and twice as heavy – but, nevertheless, completely reliable. I also observed that other American companies were going into joint ventures with their European counterparts. I told my bosses back home that we had better get into this act, or European business would be lost to us. They said they only wanted to take money out of Europe, not pump any money in. I thought it was the wrong decision and told them so, to no avail. I still think they were wrong!

The result of this decision was not long in coming. As soon as the NATO Hawk motor production run – for which we got a fixed fee for every unit - ended, the black ink on the Paris office books immediately turned to red. I said to my wife, "The new head man at Aerojet won't notice the inconsequential bleeding for at least a year, so relax." As usual in matters of a cosmic nature, I was dead wrong. My Boss came over, and at a sad meeting at the great seafood restaurant in our neighborhood, Jarasse, he told me they had decided to conduct their European business directly from home, after I had closed down the office. We had 6 months – enough time for the kids to finish their school programs. We would have loved to stay longer in Europe, but I just didn't have enough savvy to find such a permanent well-paying job there.

On the other hand, the 15 months we spent in France were a revelation for all of us! I got to travel to all the Western countries - England, Germany, Italy, Belgium, Holland, Spain, as well as Morocco and Algeria and many French cities. Our kids also got around. Our Daughter went to the Sorbonne; our oldest son to the American High School near Paris; our middle son to the Ecole Bilangue which did one semester in English and one in French; and our youngest son, 3, had a crack at a Maternelle, the French equivalent of Kindergarden. The two older kids got to travel round the continent and North Africa – a lot by hitch-hiking, which was a norm then. All but the youngest greatly enjoyed the experience. And my wife learned how to cope with the intricacies of dealing with French stores and markets and with a fractious Au Pere who loosely watched over the kids. And I learned to appreciate the quality of the education of the French engineers that I worked with.

I got no Corporate help in finding a new position back at Aerojet in Azusa. By long distance, I contacted an old friend from my Convair/ Pomona guided missile days who had just assumed managership of the very big, very important "Early Warning Satellite" program. He offered me the job of managing the test program wherein the first sensor system that we developed for the Air Force would be accepted for delivery to our partners, TRW; who were responsible for the rest of the satellite. The story of the near disaster that almost took place during the final system testing is covered in "*On the Cutting Edge*". It was a harrowing and bitter experience for "the new boy from France".

So, when I got a inquiry call from old Sandia friend Paul Rowe, who was now working at Aerojet's liquid rocket plant in Sacramento, about a possible change in jobs, I was amenable to listen, although I felt I had zero chance at getting it. He told me that his Alma Mater, Iowa State University, had put out a call to their alumni to help in the search for someone from industry to become a professor and Head of their Aerospace Engineering Department. Would I be interested? I had been an Instructor at New York University, and in the early '60s had taught graduate courses for the UCLA Engineering Extension organization. I enjoyed teaching and liked the idea of becoming a real Professor. But, Iowa! - right "after they'd seen Paree?" On the plus side, I remembered the considerate and sincere generosity of Midwesterners from my days of fighting WWII while stationed at Navy Pier in Chicago.

I discussed the proposition with my wife and family. My Mother never did know whether I would be going to go to Ohio or Iowa. It seemed like such a long shot that the family said, "What the hell, give it a go"; all of realizing that the chances that I would get the appointment were small. Surprisingly, after a visit from the ISU Dean of Engineering at our home in Claremont, we were invited to Ames for an interview. Even more surprising, I got the job!

Chapter 4
LA VIE ACADEME

- SPACE A'BORNING IN IOWA
- A ROCKET RACKET
- OVERCOMING FEAR AND TREMBLING
- REVEILLE AT REVELLE
- LEAVING TENURE BEHIND

SPACE A'BORNING IN IOWA

Our move to Iowa (1971-80) was quite a culture shock, both socially and career-wise. About the first, we had the initial feeling that in one fell swoop, our arrival had doubled the Jewish and Democratic population of the state of Iowa. We soon found that not be strictly true, but did find that were no discernible prejudices in a State where minorities of all varieties were so small in number that they were easily welcomed into the community. Politically, we quickly allied ourselves to then Sen. Harold Hughes and his equally Democratic successor, John Culver, and helped Democratic Sen. Tom Harkin win his first congressional seat. Several other stories from my life in Iowa may be found in "On the Cutting Edge"

I soon discovered that academia was the antithesis of industry. Every faculty member was a star (or so they thought) and even though I was undisputed and Department Head-for-life (as opposed to Department Chair, a 2-5 year renewable appointment) the only real hold I had over my faculty was in the annual doling out of the meager salary increases. In industry, when I said, "this way", everybody followed. At faculty meetings, when I *suggested*, "this way", only about half of the faculty went along every time. Every important issue, even unto Motherhood, would result in a tie or equally inconclusive vote. Pretty soon, I found that I was consistently being overruled by the stifling democracy that marks academia. So, then I changed my strategy:

Upon introducing an issue for discussion and action, I stated that the 'consensus', gathered from private conversations, was exactly what I wanted to happen. Pretty soon, I found that this was the kiss of death –so, without blinking an eye, I said that the consensus was exactly NOT what I wanted. From then on, the day was mine! With this strategy, I had them going my way without their knowing that they were being had. I no longer suffered from the migraines that marked my earlier faculty meeting outcomes.

My outside activities consisted of joining Rotary – to meet Ames people from all walks of life other than faculty; being active in the Orchestra Festival events which brought all the world's great symphony orchestras into our acoustically outstanding concert hall; being active in city & state politics; in sailing in the 16 foot 'Demon' sailboat that we bought in Kansas City on our way to our new home; and in inaugurating my illustrated New Orleans and Dixieland Jazz lecture series, "*Serenades for Mouldy Figges*", which are described in my fourth book. "*The World in a Jug*".

My wife, Pat, also broke new ground. She served hard liquor, previously unheard of in the Bible Belt, at the traditional annual year-opening Department Party for our faculty and grad students. She also offered non-lethal punch, which my staunch Baptist Dean preferred, nicely telling her that "he was driving". She soon was the star of the local Little Theater group: her portrayal of the raunchy Elizabeth Taylor lead role in "Who's Afraid of Virginia Woolf" being almost too much for the staid Iowans. We loved Iowa but hated the winters. Just recently (Fall, 2009), the AARP rated Ames the 5th of five best places for retirement. Were it nor for those bitter cold spells, Mrs. Lincoln, we would rate it #1!

A ROCKET RACKET

When I came to Iowa State to head up the Department, Aerospace Engineering had recently changed its name from 'Aeronautical' Engineering to accommodate the trend of the times, but its faculty was still 95% dedicated to aeronautics. Only one Astronautical course, *Orbital Mechanics*, was then included in the curriculum. I tried to balance the scales and soon introduced new Astro courses in 'Spacecraft System Design' and in 'Techniques of Remote Sensing from Space'. To my knowledge these were the first such courses at the undergrad level ever offered for academic credit in the country (the World?). I also attempted to add astronautics content in the laboratory program to complement the strong wind tunnel and aeronautics emphasis that pervaded the established Lab program.

Just before the first lab demonstration, we sent out postcards to every house in the immediate neighborhood of our makeshift firing range. The message was, "At or around 10:30 a.m. on Tuesday, April 5 (circa mid-70s), we will simulate a rocket engine firing. You will hear a loud "bang" followed by a 10 second deep rumble. Do not be alarmed, it is just a non-flammable rocket engine firing test." We also arranged for local press, radio, and TV announcements covering the coming event and even posted notices at strategic places in the vicinity of the test site – bucolic neighborhoods of individual dwellings. We had obtained the cooperation and blessing of the campus security force. We were fat, dumb, and happy.

However, when the first such actual event occurred, all hell broke loose! The phone company, police headquarters and university administration were all swamped with worried inquiries. I'm sure evacuation was contemplated by those who lived nearest to the nuclear research reactor on whose grounds we conducted the firing. True, the noise was loud -very loud - louder than I remembered the real thing being, although I was normally sheltered in an enclosed block house during a live firing. A couple of days later, the Dean of Engineering, obviously under fire from the President's office, asked me if such demonstrations were really necessary. I had busted an arm and a leg to bring this off, so I unhesitatingly said to Dean Boylan, "Dave, yes it is. It's as close to a real operation as our students will ever get, unless feature to our lab course curriculum from now on.

AEROBEE ENGINE TEST STAND

After I asked them, my former customer sounding rocket friends at NASA very kindly had shipped us a marvelous, robust, massive but still portable, test firing stand on which was mounted one Aerobee 5000 pound thrust rocket engine along with a fuel tank, an oxidizer tank, and a pressurant tank, with all the proper plumbing and instrumentation in place. In addition, they supplied key spare and replaceable parts. The Aerobee normally used high pressure gaseous Helium as a pressurant to force the fuel and oxidizer liquid propellants into the engine's thrust chamber, where they would combust. Since Helium was hard to come by, we used the more readily available and inexpensive Nitrogen gas. We, of course, would not use real propellants, since they were both toxic and otherwise dangerous to handle, although we would demonstrate propellant loading techniques dressed in the special moon-mission-like clothing that NASA also sent us. We would use water in place of the fuel and oxidizer and would conduct what is not unreasonably called "water expulsion testing".

Since I had forgotten most of the protocol and voo doo-isms that precede a hot firing, I asked for and received help from an experienced NASA engineer from the Washington area to guide us through the first propellant loading, system arming, and firing exercise. We had found an ideal "cut" in a small valley on the grounds of the university's atomic research reactor site with an abandoned cement block building adjacent. We poured a concrete pad for the test stand at the bottom of an upward slope. We figured and hoped that most of the noise would be reflected upward into the sky. The rugged concrete block building shielded the faculty and students during the firing. We rigged a strategically placed mirror for viewing and remote movie cameras for recording. Measurements of pressure, thrust, temperature and time were recorded during the test run. It turned out to be an impressive, albeit, loud exercise. Both the students and the faculty loved it – only the townspeople objected.

I have little doubt that our lab demonstrations employed the highest thrust rocket firing, real or simulated, ever conducted at a university. We repeated the event the next two Fall quarters, without outside help and with few problems. But, the effort to yearly mount the experiment was a large one, though the students always looked forward to it. When I left ISU to return to industry, the test was quietly dropped from the lab course curriculum. Worse, when I visited the campus a few years ago, the Professor in charge of the Department's lab program asked me if I could find a nice home for the test stand and other Aerobee tankage that they no longer had room for. I posed the offer to old time Aerojet rocketeer Bob Truax; he of sea launch fame, who I believe considered taking it under his wing.

Best, for me, however, was a fall out. I reported on the ISU rocket firing program at an International Aerospace Federation (IAF) meeting in Rome in 1981, and my pitch was heard by Professor Ya'acov Timnat, a noted Israeli rocket engineer and author. Several years later, when I applied for a Visiting Professorship at his university, the TECHNION in Haifa, he remembered my talk and urged my appointment. Subsequently, when we arrived in Haifa the first time in 1989, he was my sponsor. We have remained friends until his death in the late 90s; and he has written several books on rocket propulsion.

As I write this story, I am reminded of the supernumerary who was given a small speaking role, "Hark, I hear the cannon!" in an opera featuring the back ground music of the "1812 Overture". When the first cannon roar came booming out, he shouted in alarm, "What the hell was that?"

LA VIE ACADEME

OVERCOMING FEAR AND TREMBLING

In the academic year 1978-79, I took a Faculty Improvement Leave from Iowa State University to work as a 'consulting' ranch hand at the prestigious Hughes Aircraft Company in El Segundo and moved the family to nearby Hermosa Beach. My success on this venture inaugurated a subsequent program of Hughes bringing in visiting professors yearly; with HAC paying their university salary. Having been away from industry for seven years, I approached this experience with some trepidation. The fear of technical obsolescence in the aerospace business is a real one. I found myself surrounded by bright young people who were routinely applying advanced techniques to ongoing projects. In fact, it soon turned out that their only shortcoming was age; the lack of a Proustian déjà vu which can only be acquired with complementary creeping senility. Luck, plus memory, soon provided the scenario for one of life's little triumphs: the opportunity to "do well while doing good".

"Jerry, you can't do that", I said to Dr. Jerry Dutcher, an Iowa State grad who was the chief engineer of Hughes Aircraft's new LEASAT communications satellite, under development for the Navy. "Can't do what?", he answered. "Can't change the tankage design from conispherical to spherical", I answered. "Why the hell not?" " Because, the damned engine will blow up, that's why". "You're nuts – OK, tell me about it".

I was assigned to the LEASAT program office for a period of 2-3 months before my next 'rotation'. Jerry had asked me to look into the problem of whether the satellite, after being released in an Earthward direction from the Space Shuttle in low earth orbit, would remain in the sun's shadow of the 'upside-down tail-first--flying Shuttle for any significant length of time. If this were true, then the solar cells that were providing power to the satellite would not work, and the battery would be drained. I was very pleased with myself, for - by using the very same equations that I taught in my spacecraft design course at ISU - I had shown that shadowing was indeed a problem. However, it was one that could be easily solved by pushing the satellite out of the Shuttle's cargo bay at a small bank angle so that it would quickly get into full sunlight. This exercise brought great comfort to me. I had done something useful and proved that I could play with the youngster hot shots! My confidence was high.

I had just reported these results to Jerry, when the opening repartee above ensued. That very morning, I had noticed that a major design change in the LEASAT propulsion system had been made the day before. The large tanks which provided the propellants which would rocket the spacecraft from an elliptical orbit to the 22,000 mile circular 'stationary' orbit from which it would operate had changed from Hughes' usual tear-drop shaped conispheres (traffic-cone shaped with a dome at the base end At that time all HAC satellites were spin stabilized and utilized centripetal force to feed the propellants from the small end of the conisphere into the fuel lines) **to a never tried spherical shape. This had been done to increase the propellant-carrying capacity without increasing the satellite's overall size.**

On seeing the change, my mind immediately reverted to my early 1959 experience at Aerojet when I witnessed the explosion of several rocket engines right after they lit off. Vanguard, which would have beaten Sputnik by a month as first to orbit, was such a victim. Its second stage, built by Aerojet, had its engine explode during the launch process; and it took some time and more explosions before we determined the problem. The engines were all fed by cylindrical propellant tanks whose ends were spherical. Painfully, it was learned that the problem was caused by the formation of a vortex in the propellant which formed as soon as the propellant started flowing into the engine. The vortex was akin to the ones that form as bathtubs drain. It interrupted the flow of the propellants and caused fuel mixtures that blew up rather than burned. The solution to the problem was both simple and elegant. Anti-swirl baffles were welded to the tank bottoms over the propellant feed outlet, and the problem went away.

Because Hughes had always used conical-ended tanks where the propellants exited, they had never seen the problem. When I told Jerry about it and suggested adding baffles, he hesitated, using the added expense as an excuse to drag his feet; although I felt he wasn't believing a word I said. He discussed my finding and baffle suggestion with several key program and rocket engine specialists. My suggestion was not well received by these young Turks who had such an enviable record of success. I pointed out to them that they had never encountered such high propellant rate flows and tank shapes before. But, the engineering world has a built-in prejudice against the narration of historical happenings and outlanders telling them their business. This is scientifically called the NIH (Not-Invented-Here) syndrome, and is very difficult to overcome.

To his credit, Jerry finally took me to the Program Manager so that I could tell him my fears. He, Dr. David Braverman, having again consulted with Hughes Propulsion group experts (who had quickly poo-poohed my 'theory'), said that this was a good opportunity for some real engineering work, and challenged me to prove my contention. I'm sure he thought that this was a clever ruse to get me out of his hair while protecting his ass should anything go wrong.

Thus began a month of intensive work in the library and with my computer. I summoned up the wisdom contained in ancient papers on vortical flow which had been done in the late 50's, and gradually could build a case that supported what they called my 'theory', but which I knew to be the actual truth. After all, I had seen it with my own eyes; but, alas, they hadn't. I wrote a large report and gave several lectures on the phenomena to Doubting Thomas'. When the smoke cleared, I had won! Baffles were added to the design and I moved on to the next assignment with a great sense of accomplishment and the certain feeling that I could return to industry and hold my own in a non-supervisory capacity if I wanted to. I would be able to compete as a pure technician, and could forego the hand-holding management type jobs that I had previously held, but now eschewed.

Almost a year later, one of the doubters in the HAC Propulsion Group with whom I had become friendly sent me the results of a very complicated set of tests they had run to try to vindicate their contention that a vortex could only form when the tank was almost empty. It was an ingeniously designed test, and simulated the true zero gravity conditions. Just as I had predicted, the tests clearly showed the formation of a vortex all the way from the top of an almost full tank. Hughes later sent me a letter of commendation for this work. In truth, it was my finest purely technical hour.

REVEILLE AT REVELLE

We enjoyed being back in California, living on the beach where the weather was consistently good and the sailing year round. We were living in Hermosa Beach, while I was working at Hughes Space and Communications. Half way through my sabbatical year from ISU in the late '70s, I stumbled across an intriguing job opening in the academic world via the 'bible' journal "Chronicles of Higher Education". I noted the advertisement for the job of Provost of Revelle College; the most prestigious of the four colleges that then comprised the University of California at San Diego. I immediately called my friend who was Chairman of the Applied Mechanics and Engineering Science (AMES) Department and asked about the job. Since I would be a member of his faculty if I were selected, I needed to know if I was acceptable to him. He answered affirmatively and enthusiastically, and on that basis, I applied for the position. I asked several friends in high places to write nice recommendation letters for me, which they did effectively.

Things moved ahead quickly and swimmingly! Soon, I was informed that I was 'one of the candidates under further consideration'. On that basis, we started looking for places to live, favoring the somewhat honky-tonk Mission Beach area; a peninsula with access to the ocean on one side and Mission Bay on the other side a little over a block away. We were excited! Even more so, when I received a formal invitation for a personal interview a few days hence, and was informally told that I was #1 - out of 3 - who would be interviewed! Then, on the Friday before the scheduled Monday/Tuesday interviews, I got an ominous phone call. It led to this letter of March 13, 1979, sent to those who had written letters on my behalf:

"**Dear Friends:**

A funny thing happened to me on the way to U.C. San Diego! Thanks in part to your fine letters of recommendation (for which you have my heartfelt thanks), I made the "Top 3" (indeed, the head of the search committee told me I would be rated #1 if I did not stink up the interviews). A two day program of interviews was set up for March 12 & 13.
At the eleventh hour, on Friday morning, the search committee chairwoman called with the disastrous news that I was not acceptable to "some" of the faculty members of the Department to which I would be appointed. The basis

of their disaffection was lack of recent scientific papers (my 'classified' papers did not count). This was shocking, since the first thing I did before I applied for the Provost's job was to discuss my joining his faculty with its Department Head, an old acquaintance. He said he would be delighted to have me apply – but apparently did not clear this with his faculty until he was told I would be invited down for an interview. Although this may seem strange to you, since I would do very little teaching (2 courses per year), I think that the faculty which is composed of famous long hairs, simply did not think that I was in their scientific class (which I can not deny, being more of an engineer than a scientist) and were fearful that I would one day 'step down' as Provost – and that they would then be stuck with me. My counter was the assurance that I would resign rather than impress myself on them if they did not want me. As of this date, that ploy did not work, and I am writing it off – honored that I got as far as I did.

Naturally, I am bitterly disappointed, as it was undeniably a good job (albeit, the salary was not quite as much as I am presently making at ISU). However, I will depend on my normal sunny nature to pull me through – and I suppose Pat will really not divorce me and will go back to her corn planting.

Again, my thanks for a valiant try. Because of you, I know that 'Someone out there likes me!' "

Later that year, at our Technical Society *Space Systems Committee* meeting, I talked hypothetically to a fellow committee member who held a high position at TRW in Redondo Beach. He told me that if, after I returned to Iowa, I still wanted to come back to industry, I should let him know and he might be able to offer me a job. Halfway through the next academic year, I did and he did. As a family, we weighed the options and decided to return West. It turned out to be a good decision – and finally got me back to the beloved beach.

LEAVING TENURE BEHIND

The ocean – any ocean – has always been a great attraction to me. I like sitting on the beach and riding the salty waves. I like sailing in it – it doesn't play tricks like sudden wind shifts and gusts that rise up in the inland lakes. I guess I learned love it in my youth when we yearly spent a month of summer at Atlantic City. When we lived inland in Claremont, we would always take our summer vacation at Newport Beach. In 1958, I had a chance to work there, but turned it down as too risky career-wise for my growing family. So, when the chance to live by the sea came again in 1980, with family blessing, I grabbed it. The South Bay part of Los Angeles, where all the space action was taking place, seemed like a little bit of heaven. We moved there that year. And, except for a year of living in Israel, we have been here ever since; and the next Chapters have stories from this final era in my engineering life and retirement years.

The late '70s sabbatical experience at Hughes (now, Boeing) had been exhilarating. I realized how much I missed the excitement and competition of industry. We also realized how much we missed California and year-round sailing, and how nice the South Bay area was. I discreetly inquired about a regular job at Hughes, but they said that they had now initiated an annual 'Visiting Professor' program - so pleased were they by their experience in my just completed year - and that hiring me would make the new program look like a recruitment ploy. I scratched around and found a good 'Senior Systems Engineer' job at their arch rivals; equally noted and close-by TRW (now, Northrop

Grumman). I started working there in August, 1980. I knew I had the option to go back to Hughes after a decent interval, but it turned out that I enjoyed the TRW action, and stayed there until I retired from full time industry work in 1988. I also had the opportunity to resume teaching my space courses at night at the University of Southern California, which I did starting in 1982 until I retired from there in 1996.

In August, 1980, we moved to a Town House in Hermosa Beach, two blocks from the beach and 4 blocks from King Harbor Marina. Soon, I was co-partner in a Catalina 27 sailboat and began sailing twice a week, a ritual I maintain to this day. In leaving Iowa State, I relinquished a tenured position – a guaranteed job for life until retirement. In the time frame 2009 economy, this might be looked on as a very foolhardy thing to do. However, at that time -1980-, the aerospace industry was healthy and vigorous, and I was filled with confidence gained from my sabbatical year experience. I had enjoyed teaching a great deal, and meant to continue doing so. Fortunately, the Chairman of the USC Aerospace Engineering Department, knowing me and wanting to expand his Department's mostly aeronautical course offerings more into astronautics, asked me to bring my academic astronautical sideshow into his stable at the graduate level. We agreed that, my boss at TRW willing (which he was), I would teach a course a semester in the evening, thus not interfering with work.

Chapter 5

SPACE BY THE SEA

- **LIFE AT TRW**
- **ASTRONAUTICS AT USC**
- **BEATING A DEAD HORSE**
- **A SMASHEROO - WAITING TO HAPPEN**
- **SOLAR SAILING AT THE TECHNION**

LIFE AT TRW

Not many tenured professors – with a lifetime of employment assured as a result of being tenured – abandon academia to take a chance with the vagaries of the industrial world. I, on the other hand, having accomplished most of the goals I had set out for the Aero. Department at ISU, and having seen, via my sabbatical year at Hughes, that the California climate and lifestyle it afforded was hard to beat, had the desire, strength and bargaining power to leave a 'for life' position and people that I and my family were very fond of. My sabbatical experience at Hughes had shown me that I could compete with the younger very bright engineers working at the trade. So, I took the chance at the brass ring.

With new found confidence that I could make it back in the world of Industry resulting from my technically successful sabbatical year, '78-'79, at Hughes, I attacked my new assignments at TRW with great enthusiasm. At the outset, I was put in charge of a research project, which turned into a study contract, the Shuttle Bus, whose story was told in "*Cut-*

ting Edge". I next led a winning proposal on a Mars explorer satellite and completed the study.

Then, spotted by a Vice-President who would become my next boss, I took on, kicking and screaming - for I liked 'Line' as opposed to 'Staff' positions - a prestigious and better paying high Group level staff job; Director of Technology Planning. I hated to leave the trenches, but the inducements were more than I could refuse. This change put me in charge of approving and overseeing all company-sponsored research as a 'side' job – my main function being to somehow make sure that TRW had all the advanced research developments around the world covered and tracked, so that we could never be surprised by another outfit having technology that we were not aware of or could not better. This was a cushy job that came with a secretary and a couch in a corner windowed office. It turned out, however, that the main assignment could not really be done. There are simply too many vagaries in the world of scientific research to be able to predict when and if a break through will be made. I wrote and presented a paper to this effect, and gave a copy to the BIG BOSS. There went the cushy job!

I did other things to keep me busy: I started teaching the two courses that I had developed at Iowa State: "*Spacecraft Systems Design*" and "*Principles and Techniques of Remote Sensing*" at the University of Southern California, where I was appointed a full professor. This, in effect started the Astronautics option in the heretofore airplane-oriented Department of Aerospace Engineering. Further, I cut a deal with my boss in which he agreed to send me to the International Astronautics Federation meeting every other year so long as I had a technical paper accepted. This got us to some very nice places like Rome, Innsbruck, and Lausanne. I served on two high ranking national Technical Committees of my technical society, the 'Space

Systems' and 'Space Transportation' committees and also as the Chairman of its Los Angeles Section. After I left the Staff job, I got back into proposal writing, my true forte, and to being the systems engineer on advanced study programs - until I retired. I enjoyed being a 'ranch hand' again. Staff positions, albeit very important, lack the excitement of competition.

While I was on Staff, I got an assignment that nobody wanted. It was to head up a committee (AAP – Affirmative Action Policy) to oversee the Group Staff's response to the rather rigid government-imposed rules concerning a new dictum – Affirmative Action. If you didn't establish quotas for minority hiring and show that you were actively attempting to meet these quotas, the government could take reprisals; up to not granting you contracts that you had won technically. In truth, I felt great sympathy for the program's goals and most of the people on my committee were activists of like mind..

But it was not all grim bean-counting and hiring goal-establishing. We met weekly, setting personnel 'quotas' (although it was death to call them as such) for hiring and promoting of minorities. It was interesting - half serious and half boondoggle. Below is an example of the latter, which permitted a synergistic melding of work and avocation:

Interoffice Correspondence
TRW Space & Technology Group

Subject   *An Offer You Can't Refuse*
From   *RFB your CAPO*
Date   *February 21, 1984*
Group Staff   **AAP Committee**
Location/phone   **R5/1031 61824**

At great savings to the women and minorities of the world, we have chartered the fabled sloop, "Poulet de la Mer" and its equally noted British Captain, R. Foxroy Broadbeam, for a series of late Wednesday afternoon group staff AAP committee meetings. Alas, the sloop, a magnificent Catalina 27, only holds six committee persons comfortably, so a number of cruises will be necessary to get the work of the committee done. Reservations will be taken by contacting the Captain's executive officer, Cheryl, at x61824. The boat leaves from slip B-9 at the King Harbor Marina (see map) promptly at the arranged times. Tennis shoes are de rigueur. The Captain provides ice and a grim visage - you bring the committee business!

It is anticipated that the initial meeting will take place on March 21; then April 4, 11, 18, 25, etc. Make your reservations now!

In my 63rd year, the tempo of action at TRW started slowing down. Government money was tight and space vehicle procurement was slowed. Company belts were being tightened. My assignments became less interesting and exciting. I felt I was being pushed towards early retirement. Indeed bonuses were being offered to those old timers who opted for 'early-out'. Financially, retirement was entirely feasible. As a result of inheritances from my Mother and my former Nurse, as well as retirement packages from Aerojet, Iowa State, and now, TRW, our monthly income, supplemented by Social Security and IRA accounts, would remain about the same.

So, given a bitter but helpful 'push' by management, I left my beloved industry in July 1988. Getting into the retirement mode was not the problem for me that it was and still is for some of my buddies. Mostly, I missed the camaraderie of my friends at work and the mental stimulation of tackling a new problem or writing a new proposal. I continued teaching at USC one night a week for both semesters until

1996. In early retirement, I occasionally helped my old TRW friend, Dr. Jim Wertz start his new company, Microcosm, Inc., which is still in operation.

I began a daily regimen of writing in the morning and evening, taking a daily hot tub/ swim followed by a walk, and sailing twice a week. After some rumination, I began the legal action featured in this chapter as a hobby, not for revenge or money. I continued, as I still do, to help out with my Technical Society's activities, and occasionally review the talks of dinner meeting speakers for their Newsletter. I dare say the hardest adjustment was my wife's - what with me being around the house every day. We also started traveling a lot. All this continued to keep me off the street.

ASTRONAUTICS AT USC

My timing viz a viz college athletic prowess was pitiful. During my tenure at Iowa State, the football teams were consistently mediocre and didn't improve until after we moved back west. On the other hand, my new allegiance to the University of Southern California Trojans promised for better days – for they had been a national powerhouse all the time we were freezing in Iowa. The on-going story was that the football team longed for a faculty that they could be proud of. Nevertheless, during the 14 years that I taught there – 1982 to 1996 – they never fielded a national power. As I first began to write this, seven years after my retirement from USC, they were the number one football team in the country. But – and this I am proud of because I started it – I now think they are the best school, #1, in the world to get an education in Astronautics, at both the graduate and undergraduate levels. How did I get a professorship at USC while holding a full time day job at TRW? Simple – the courses I taught were presented one night a week from 6:30 p.m. to 9:15 p.m.

CATCH A ROCKET PLANE

C. L. Max Nikias
President

Robert C. Packard
President's Chair

Malcolm R. Currie Chair
in Technology
and the Humanities

February 3, 2012

Dear Bob,

On behalf of the entire University of Southern California community, I want to thank you warmly for your stellar service to the Viterbi School of Engineering, and congratulate you on your many professional achievements. You stand as a true pioneer in your field, and your exceptional work as a scientist, engineer, and teacher has had a profound impact on astronautics engineering. You helped build USC into the world-class institution it is today, and for this we will always be grateful.

Over the course of your career, you have led innovation in spacecraft design, as well as space engineering education. Your dedicated efforts to develop more robust astronautics programs have immeasurably advanced engineering education in this country. At USC, you remain highly esteemed for your vital work that led to the creation of an independent Department of Astronautical Engineering. I know that your leadership, vision, and expertise have greatly benefited your students and colleagues, and through them, your wisdom will continue to touch our society.

We feel extremely proud to count you as a member of our Trojan Family, Bob. We extend our heartfelt gratitude to you, and wish you continued success and joy. We hope to see you on campus often in the coming years!

Yours truly,

Max

C. L. Max Nikias
President

'Astronautics'- the study of outer Space - is a word that is not as deeply ingrained in the national consciousness as its sister sport, 'Aeronautics'- the study of atmospheric vehicles. One of the reasons for this is historical: Aviation dates from the early 1900's, while Astronautics started circa 1957. Even as I wrote this, I noted that my Windows '98 program underlined-in-red the leading word in this paragraph, connoting 'astronautics' was then not recognized by the pea – sized internal dictionary. But, I grew up with Astronautics and helped make it the legitimate separate category academic degree program it is today in several schools. I always believed it was a new legitimate branch of engineering and fought for this belief.

The road to achieving this academic legitimacy for Astronautics was long and hard, impeded chiefly by those I called "The Battleship Admirals" of the aerospace profession. Who were they? Why, they were the aeronautical faculty who had spent a great part of their careers teaching and researching in the old fashioned world of AIRY-planes and MISS-iles. They didn't care to see a new field siphoning students and money away from their academic playgrounds. They didn't even think it belonged in the purview of Aerospace Engineering Departments, since there wasn't much fluid flow, and lift and drag, involved. But, as with everything, truth and righteousness will win out. Which gets us to today's situation:

It is my contention that a student, either an undergraduate or a graduate student, cannot get a better education in Astro than is now being provided by Dear Old University of Southern California (or in football, too, for that matter at least in 2003-8 period). I say this with some pride, because I essentially started the program in 1982 when, at the invitation of the then Department Head, I began teaching my 1972-invented Iowa State-imported courses, "*Spacecraft Systems Design*" and "*Principle and Techniques of Remote Sensing*" in the Fall and Spring semesters. Prior to then, as was typical throughout the country, the only space related courses being taught were in Orbital Mechanics, the establishment of operations in orbit in space plus some elements of rocket technology in propulsion/engine courses.

There were two reasons for this: One, except for orbital basics, there were no space engineering textbooks that a 'lay' teacher could teach from; and, Two, there simply weren't any people in academia qualified to teach true Astro courses. I say 'true', because orbital mechanics is strictly a mathematical exercise that requires only knowledge of basic physics and mechanics. I, on the other hand, having toiled as a pioneer in the new space field in the late '50s through the '60s until I left industry to join academia, was uniquely qualified. I suspect that I was the first space knowledgeable pioneer to leave industry for academia.

Just as I had found at Iowa State, the USC student desire to learn about this new field was overwhelming and USC, being in the very center of the newly burgeoning space industry, was perfectly positioned to take a leading role. I had already developed most of the class notes needed for my two courses, now at the graduate school level. Almost at once, my spacecraft design course attracted 60-80 students mostly from the local industry and my remote sensing course about half this number. Soon, I was simulcasting my lectures, over closed circuit TV, to the local aerospace industry sites as far away as Arizona. Later, I simulcasted (or was taped for distribution by mail) over national TV. It became apparent to my sponsoring Aerospace Engineering Department and the College of Engineering that the program should be expanded. Being in the center of government/industry space endeavors, with Boeing/Hughes, Northrop-Grumman/TRW, JPL (Jet Propulsion Laboratory), Raytheon, the US Air Force and Aerospace Corporation, and a flock of smaller companies also in the neighborhood, Adjunct Professors, highly expert in everything spacial, have been easy to find and are at the top of their game. And, fighting against the resolute majority fluid flow faculty, some space-oriented new faculty were hired.

One of these, Professor Mike Gruntman, grew up with the Soviet space program but succeeded in escaping the "Communist Paradise" (his words) as soon as the opportunity offered itself. While he was young, his father was the Chief Engineer, responsible for the construction of and early operations of the Baikonur space port, still Russia's main launch facility. In a gradual transition, Mike took over my spacecraft design course and I developed a JPL engineer to take over my remote sensing course when I felt that I'd better retire since, not working anymore, I could not keep up with the scientific and engineering advancements.

Mike, a person of great energy and persistence, started putting together a program of graduate and undergraduate classes that utilized both regular and adjunct faculty. The student population of the Department started swinging heavily in numbers towards Astro as opposed to Aero. Finally, Mike was able to establish an undergraduate degree option in Astronautics as well as Master's degree graduate program along with a Certificate Program in Astronautics. To my mind, the program is the best in the country and maybe the world, insofar as pure education in space topics is concerned. It is being well advertised in the magazines and by learned papers. The program is detailed in some of the illustrations that follow.

The struggle was not for naught, but internecine warfare continues to this day as the aeronautical admirals refuse to go down with their ship. Peaceful co-existent in a cold war atmosphere will probably go as long as they keep making air-going airplanes and missiles. But, Space has arrived in academia – it's here to stay.

In fact, in April, 2009, I received the triumphal email below from Mike that the new president of USC, formerly the Dean of Engineering who had originally established the 'Astronautics and Space Division' in 2004, had now approved the school's decision that as of July, 2010 there would be a new Department of Astronautical Engineering!!! The 'old' Aerospace Engineering Department, where I had started the space program, had now merged with Mechanical Engineering.

13 August 2004

Astronautics Adjunct Faculty and Lecturers and Program Friends

---

Dear Colleague:

In order to position the USC Viterbi School of Engineering to take full advantage of rapidly growing opportunities in space, Dean of Engineering Prof. Max Nikias announced today the creation of a new Astronautics and Space Technology Division (ASTD).

ASTD will be an independent academic unit within the USC Viterbi School of Engineering and function in a manner similar to an -academic department. The division will be governed by the same -rules and policies that apply to academic departments, with its own budget, faculty self-government, and representation at the School's committees and other bodies. I have been appointed chair of ASTD effective August 15, 2004. Aerospace engineering faculty,

———————   ———————   SIGNED   M. GRUNTMAN-

A TRIUMPH AFTER MANY LONGS YEARS OF STRUGGLE!

The final nail in the coffin - spacemen vs. battleship admirals:

From Michael Gruntman <mikeg@usc.edu>
Subject USC Astronautics: mission accomplished - the new journey begins

22 April 2010

To Friends and Supporters of USC Astronautics

Astronautics at USC has reached a major milestone.

The Dean of Engineering has announced transition – effective July 1, 2010 - of the Astronautical Engineering Division into the Astronautical Engineering Department.

In 2004, the University of Southern California (USC) established - in a bold experiment - a new academic unit, the Astronautics and Space Technology Division, operating as a department in the USC Viterbi School of Engineering. The Division focused on space engineering, in contrast to common arrangements in the American academia where astronautics and aeronautics are combined in aerospace departments and programs. The Division developed the full set of degree programs (Bachelor, Master, Engineer, PhD, Graduate Certificate) in Astronautical Engineering. All programs are highly successful. The Master's program is among largest in the nation, reaching students across the United States through the Distance Education Network.

BEATING A DEAD HORSE
(WHILE TILTING AT NASA WINDMILLS)

Although I tried to get NASA interested in the concept of a space lifeboat for the International Space Station in the early '90s, they continued undeterred on their inevitable path of using crew escape systems as a means to justifying what they really wanted: An updated Space Shuttle. As one program after another: Crew Rescue Vehicle, Crew Return Vehicle, Crew Transfer Vehicle and, lastly, Orbital Space Plane were cancelled, they failed to see that the problems of crew escape from the ISS and the routine delivery of crew to and from the station called for two different solutions. During the '80s, I tried to gore the ox:

So began an uphill campaign against an intractable but sincere government agency apparently incapable of being light on its feet. In the pages that follow, I have listed only a few of the articles, Letters to the Editor, letters to and e-mails to NASA and influential congress persons, and to my dead sainted Mother. Not listed are personal face-to-face encounters I had with the number 2 person in NASA, and other high NASA officials at their Huntsville, Houston, and Bay Area Research Centers, the many e-mail pleadings, and a complete listing of the several technical presentations I made in the Los Angeles area. A list of what I did follows:

• A late 2002 'Correspondence' to AEROSPACE AMERICA, the monthly magazine of my technical society again posing the need for a 'throw-away' space lifeboat to solve NASA's ISS crew escape problem.
• A 'Letter to the Editor" of the South Bay DAILY BREEZE after the Columbia tragedy suggesting that a lifeboat could be incorporated into the cargo bay of the Space Shuttle as

an escape means for Shuttles whose safe landing capability might be in doubt.
- A notice about a dinner meeting at which I was the featured speaker, talking on "Escape From The Space Station".
- An abstract of a talk I proposed to give in a technical session on space operations at "Space 2003", a national meeting of my technical society to be held in nearby Long Beach. It was not accepted, probably because it was thought to be too controversial.

When the Orbital Space Plane program was cancelled, and the new national goals in space became directed towards lunar and Mars exploration utilizing the newest darling, The CEV (Crew Exploration Vehicle) and its growth versions, I finally realized that the jig was truly up for my beautiful paraglider delusion. In spite of this epiphany, I still submitted an abstract similar to the one for 'Space 2003' to present at the annual National Space Sciences meeting to be held in Reno in January 2005. This time, however, I submitted it for presentation at a Historical, as opposed to Technical, Session. That was probably its best 'home' now. It was accepted and, at age 79 then, I almost broke the Guiness record for being the oldest-ever speaker at this meeting. Why do I do such silly things? Well, there's a great Basque restaurant in Reno that I remembered fondly from many past trips to the same meeting in days of yore.

A MODICUM OF RECOGNITION

> EXCERPTS FROM *AIR & SPACE* MAGAZINE
> "POD PEOPLE" by Jim Oberg, p.58,
> November, 2003

"- - - .Launched on a sounding rocket to an altitude of 96 miles over New Mexico, the craft dove back toward Earth at a speed of more than 5,000 mph. Being so light, it didn't generate as much heat from atmospheric friction as Glenn's capsule had, so it had only a thin coating of thermal protection – no shield. Odder still, it was inflated like a balloon in a Thanksgiving Day parade.

The contraption was called IMP, for Inflatable Micrometeoroid Paraglider. - - The actual flight of the IMP, on an Aerobee-150 sounding rocket in 1964 had mixed results. -- The idea of inflatable reentry vehicles had caught the imagination of engineers, including some outside of NASA, who started musing about light craft that could make the return trip all the way from Earth orbit. - --

One of those engineers was Robert Brodsky, who at the time was in charge of bringing in work to Aerojet-General's Space General division, which had built the Aerobee rocket. Aerojet had designed the electronics and other systems for IMP, while Goodrich had built the structure. Even before IMP's test flight, both companies started pushing the concept of inflatable reentry with their customers, --

----In the early '60's. the Pentagon was flirting with developing its own USAF astronaut program (see "A Sudden Loss of Altitude," June/July, 1998) and was interested in proposals for a small storable craft that could return a person from orbit in a hurry. "I got the idea that we could alter the IMP design sufficiently to turn it into a space lifeboat," Brodsky says. He saw it as an option for dire emergencies only, like a life raft on an ocean liner. - - "There were only two reactions," Brodsky recalls. "Initially, sheer incredulity. Then – seeing the challenge - great enthusiasm." In 1962, the Air Force Materials Lab awarded his company $250,000 to look into the concept: the funding later grew to well over $1 million. "In those days that was a lot of money," he says.

The project lead was Jesse "Bud" Keville, a 37-year old engineer who set to work designing the lifeboat and building and testing structural components. Space-General called it Project FIRST, for "Fabrication of Inflatable Reentry Structures for Test." Keville's team kept the basic IMP paraglider with its three inflatable struts, and placed a prone astronaut in the center strut. The lifeboat weighed a mere 850 pounds. Stowed on a spacecraft, it could fit in a three- foot by 10-foot cylinder; inflated, it was 23 feet long, with a wingspan of 28 feet. The engineers even came up with a deluxe three-person version, and a six person model that weighed a ton.

CATCH A ROCKET PLANE

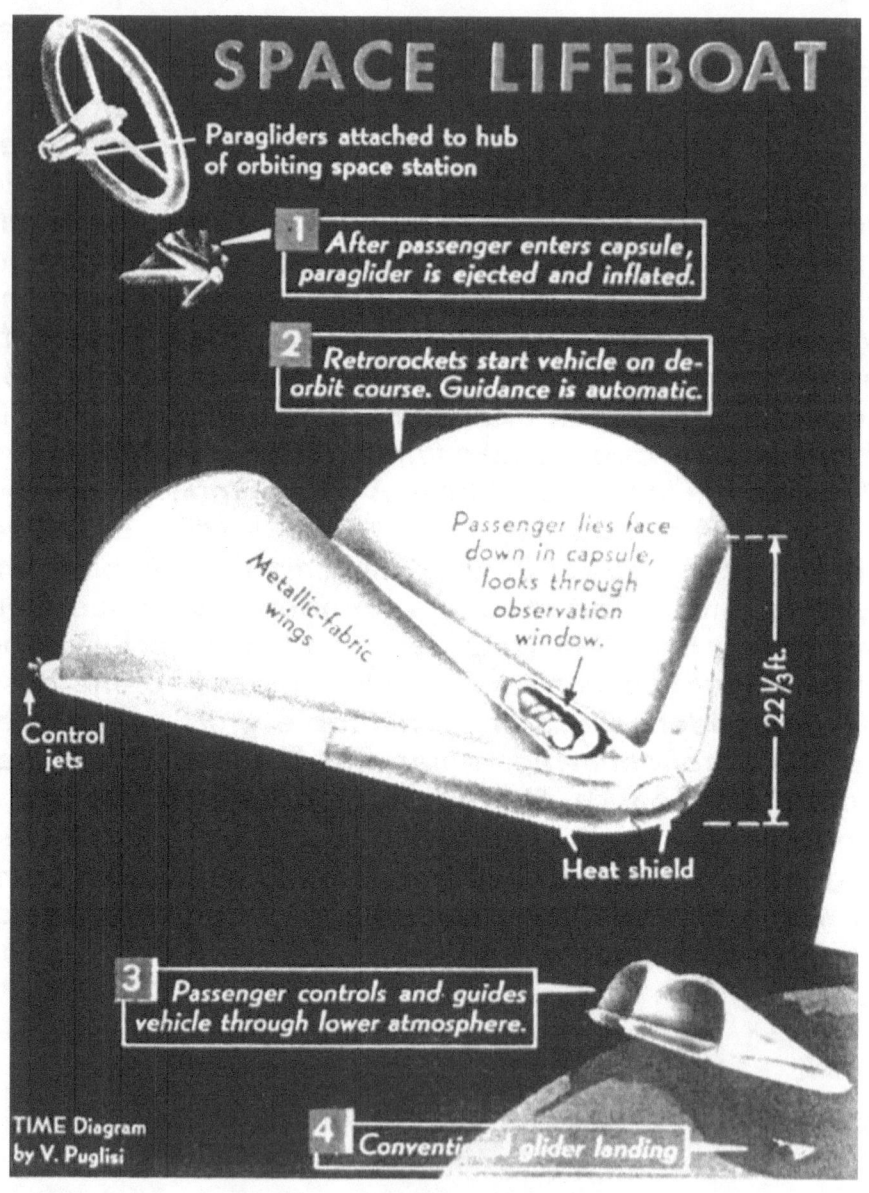

"from *TIME* magazine, Feb. 1, 1963" THE SINGLE PASSENGER VERSION WEIGHED ABOUT 1000 POUNDS INCLUDING CREW MEMBER AND PACKED INTO A CYLINDER 10 FEET LONG AND 3 FEET IN DIAMETER

The inflatable wing spars were made of nickel-chromium alloy mesh. For thermal protection, the mesh was saturated with liquid silicone, and covered with another layer of silicone rubber. A vehicle returning from orbit would experience more heating than IMP had during its suborbital flight, and vacuum chamber tests showed that this material could handle temperatures of more than 2000 deg. F.

The wing material, says Keville, "resembled a lightweight burlap." At first he had trouble finding a textile company that could handle the tricky job of weaving metal yarn, but he finally found one called Prodesco. Keville spent weeks in the Small Pennsylvania town of Perkasie, which had "only one general store and a Quaker church."

The FIRST lifeboat was designed to be folded up in a small container on the outside of a spacecraft. An Astronaut abandoning ship would enter the pod through a small hatch leading to the outside. After inflating the paraglider with Nitrogen fed through a hose or from gas bottles, the escapee would fire solid rockets attached to the closed hatch cover to de-orbit the craft. The fall from orbit (400,000 ft.) down to 120,000 ft. would take half an hour, with attitude control jets used for maneuvering. Once it became aerodynamic in the lower atmosphere, the paraglider could be steered by wing warping. The landing would take place anywhere within a footprint 450 miles wide and 1400 miles long.

Five years of research convinced the FIRST engineers that the concept was feasible. Unfortunately, by the late 1960's it was no longer wanted. Neither Apollo or Skylab, NASA's first space station, were in the market for a bailout system, and the Dept. of Defense was already starting to back away from plans for its own station. Inflatable lifeboats had become the answer to a question no one was asking.

From the beginning, Brodsky had grander things in mind for FIRST, based on the futuristic schemes that Wernher Von Braun and others were espousing at the time.

"FIRST was begun to meet an apparent need for a wheel-like rotating space station," he says. "It was terminated when it was apparent that we were too early. The idea of a large manned space station was no longer in vogue."

That hadn't stopped other engineers from exploring similar bailout concepts, though. Other companies had learned of the FIRST project, and throughout the '60's they came up with various ways to improve on it. Some aficionados of personal bailout systems still hold out hope that NASA will one day look in their direction. But so far, the Agency's plans do not appear to include one-person life-boats.

FIRST designer Robert Brodsky still talks up the idea. He accuses NASA of being close-minded on the subject of inflatable reentry vehicles. "Lack of a mission need stopped the program in the '60's," he says. "But 'Not Invented Here' is stopping it these days. Several years ago, he proposed that the Agency update the old FIRST concept as a Crew Rescue Vehicle, or CRV, for the International Space Station. "They patted me gently on the head, because their idea of a space lifeboat is radically and very expensively different from mine."

John Muratore who, until it was cancelled, managed NASA's CRV program denies that the Agency turned a blind eye to inflatables. "It's an interesting technology, and we have looked at it," he says. "The problem is that you chase a weight curve. You add weight to withstand stress, so the heating goes up, and the vehicle gets bigger and heavier." His team studied a dozen designs and did a deep literature search. They heard from the inflatable backers. They just weren't convinced.

All of which leaves Brodsky, Kendall, and the other pioneers of inflatable concepts where they've always been. Despite the technology's interesting past and its promise for the future, it still doesn't have a great present.

SPACE BY THE SEA

Here are pertinent sample rants:

LETTERS

Haven't We Tried This Before?
"Retro Rocketeers" (Apr./May 2004) failed to mention that around 1991, the Assured Crew Return Vehicle (ACRV) project office at NASA's Johnson Space Center in Houston looked into using a variation of the Apollo vehicle as an emergency return vehicle for the then-unbuilt space station. We had two contractors on board (Rockwell and Lockheed), and both were looking at capsules, with Rockwell's design resembling a scaled-up Apollo. But no one outside of our office was particularly interested, so our office disappeared.

    Dave Eichblatt
    Houston, Texas

James Oberg's "Retro Rocketeers" and his earlier "Pod People" (Oct./Nov. 2003), in which I was quoted, both deal with vehicles for making emergency escapes from the International Space Station, and both clearly point out that NASA still has not come to grips with its future program requirements and how to satisfy them.

 NASA is looking for a single solution to (1) the near-term problem of providing an escape system for more than three ISS crew members, and (2) the need for a crew exploration vehicle (CEV) for moon missions, and presumably the need to later grow that vehicle for a manned Mars mission. By looking for one solution, NASA is going down the same disastrous path that has caused the cancellation of its last three attempts to replace the space shuttle while providing vehicles for crew return from the ISS. Again, the agency has been biting off more than it can chew.

If NASA completely divorced the two, as I and others have proposed, the agency would stand a better chance of achieving all its goals. I have argued that the life-boats needed for each manned module of the ISS should have a large, un-supported land and/or sea landing foot-print, and that they should thus be relatively unsophisticated winged vehicles. These vehicles would be analogous to an ocean liner's lifeboats; i.e., their use would make success probable but not certain. Because we really don't expect emergencies on the ISS, we shouldn't employ a sophisticated, high-priced CEV in a standby mode to provide this capability. Besides, we now have two gruesome examples of highly sophisticated reentry vehicles that did not make it.

In lieu of following its usual practice--coming up with a huge in-house effort to produce a specific design - NASA should formulate general performance specifications, give them to contractors, and ask for preliminary system designs.

Then the agency should choose two or three concepts with great, potential and contract with the companies that produced them for more-detailed designs.So long as NASA carries the baggage of past designs, it cannot hope for optimum, timely, and affordable solutions to its problems.

R.F. Brodsky
Redondo Beach, California

The February 1960 issue of *Mechanics Illustrated* has a concept for a "Lifeboat for Spaceships," designed by a Frank Tinsley, that has the same operational profile as the Apollo; the reentry is virtually identical, including landing in the ocean.
More proof that what goes around, comes around.
Mel Goddard Brampton, Ontario, Canada

AIR & SPACE JUNE/JULY 2004

SPACE BY THE SEA

A SMASHEROO – WAITING TO HAPPEN
(A SPACE OATER)

Near as I can foresee, barring a miracle, my escape paraglider will always remain a mere historic monument to man's folly. While I was still in a crusading mood trying to convince NASA that they needed a space lifeboat, a beautiful scenario danced before my eyes. Wouldn't it be funny if I finally made fame and fortune by its slick application in a modern terrorist opera? The right vehicle for these troubled times? While at TRW, I had a co-worker colleague, Irv Spielberg (yes, yes, a relative). I just knew his Uncle Steven would love it!:

My earthshaking letter went like this :

" **May 26, 2003**

Steven Spielberg
c/o Dreamworks SKG
100 Universal Plaza
Bungalow 477
Universal City, CA 91608

Dear Steven:

I am not your normal crank; I am a legitimate crank – as witnessed by the enclosed TIME magazine article about my old, but newly pertinent invention. Moreover, I can be vouched for by my former TRW colleague, Irv Spielberg, whose nephew I believe you are.
In any case I have enclosed a scenario for a space 'oater' that I think might be right up your alley. It is very up-

to-date. It links terrorism with the International Space Station, and allows for a modicum of sex, lots of swashbuckling, and uses my space station escape vehicle invention as a diversion. It is "Apollo 13" brought into the near future. It requires the cooperation of NASA, which I think they would be more than glad to provide.

Should you like the idea and make plans to develop it further, I ask only two things: Reasonable compensation for the idea and hiring me on as a technical consultant; actually a wise move on your part since I am a space pioneer (see Marquis' "Who's Who in America" and the TIME mag article).

Thanks for your consideration – I think it's a winner!"

With this letter, I enclosed the following scenario:

TERROR ON THE ISS

It is 2013, and the International Space Station maintains a crew of about 10 men and women (allowing love interests to develop). A sinister terrorist organization is able to smuggle a crewperson, a long time native mole, aboard the station, and he cleverly takes control and holds the others in virtual captivity. (Alternatively – 2/3 Hijackers (who now could be 'foreign' – if you get my drift- astutely avoiding racial profiling) succeed in clandestinely getting themselves aboard the initial flight of the Orbital Space Plane crew transport vehicle (to-be-developed, now under study by 3 aerospace firms) before it is launched to re-crew the ISS, and take over once it achieves orbit.

The good controller folks on the ground have no idea that chicanery has taken place. The terrorist(s) control all communications. While in virtual captivity on the Space Station, one of the crew gets pregnant – after an illicit but precedent breaking affair aboard ship. Our hero and his lady love decide to escape in one of my two-person lifeboats in order to have her not deliver on-board and to warn the world that the terrorist(s) is/are planning a nefarious attack from space, using his engineer 'slave' laborers to build a 'device'. They have a hazardous flight (while the first part of the action needs only NASA cooperation using one of their many Space Stations mock-ups and their conception of what the OSP will look like – this escape part has to be done using modern art) and reentry in the so-far untried – in real life- 'lifeboat' paraglider, but successfully land at sea (she, now in her 5th month, gets sea sick, 'til they pick them up). The good guys, led by our now Father-hero, then organize a surprise attack using a second OSP to win back the station as the sunset covers the Earth.

COMMENT: This is 'Apollo 13' brought into the future. Get yourself some financing and a few clever script writers and you'll make millions. Divert some for me and mine. Have your people talk to my people about the details.

I see a place for Pierce Brosnan or Harrison Ford. I can only see Zeta Jones as the pregnant heroine who initiates sex-in-space – heretofore only discussed in hallowed halls. A possible villain – Gene Hackman?

(submitted by RF Brodsky exclusively to Steven Spielberg)

CAN YOU TELL ME - WHY HASN'T THIS HAPPENED?

CATCH A ROCKET PLANE

SOLAR SAILING AT THE TECHNION

As soon as I retired from TRW, my wife and I decided we would like to accelerate our travel ventures. It would take a few years to get into the mind frame of spending our own money for travel, after a lifetime of company-paid trips. I applied for a Lady Davis Visiting Professor fellowship at the TECHNION (Israel Institute of Technology) in Haifa for the Fall Semester of 1989. I offered to teach my course in "Spacecraft Systems Design". One of their Aerospace Engineering faculty knew of me, offered to be my mentor, and championed my appointment. We contacted my distant Cousin-on-my- Mother's-side, Susan, whose husband was also a Professor of Industrial Engineering there, and they eased our first entry into the old country. Haifa, the site of the school, is a beautiful beach city, as well as the major port of the country. And, best of all, it has a small boat marina for sail boats.

When I met my first class in spacecraft design, over 50 students sat in on the first lecture. A week later, only half that number showed up. I was puzzled and asked an obvious English-as-his-first-language student, "what was going on". As his story unraveled, I began to see in what a highly competitive academic environment I was now operating.

The TECHNION is to Israel what MIT and Cal Tech are to the United States. It is the premier technical school in the country and was founded shortly after the war of independence, and, with Hebrew University in Jerusalem, shares the title of being the oldest state-supported higher learning institution. Almost to a person, the industrial tycoons and leading engineers in Israel wear the old school tie. It is like a closed corporation.

The school was first located in the Hadar, or middle section, altitude-wise, of Haifa, but now is located on a large, lovely campus in the upper Carmel section of the city. At first, they were going to teach in German, as a tribute to the strong pre-war technical position that German scientists and engineers held throughout the world. But, national pride- of which there is a great abundance, and a beautiful thing to experience – won out. So, the official language of the school is Hebrew. Nowadays, however, what with the large faculty influx from the Soviet Union, the recurring joke is that Hebrew is now the second language - in favor of Russian. The students are allowed to take one course per semester in a foreign language, which is usually in English.

Their course load, both at the undergraduate and graduate levels, is barbaric by US standards, and the student competition is fierce. There are only a limited number of educational and industry/ government slots available. The best students will get the best, highest paying jobs. It's as simple as that! It is my feeling that there are a considerably higher percentage of women students than in the States. There is also the peculiarity of students being called up for two or more weeks of military service right in the middle of a semester. Although such service appears essential to the security of the State, one wonders if this is the best way to run a railroad? All students speak English, which is their normal second language. However, some are reluctant to do so, which is natural.

The Faculty are mostly ingrown, but most have received at least one higher degree in the United States or, to a lesser extent, in Europe. Most take frequent sabbaticals, again mostly in the US or England. If you listen to them, they are miserably paid, and this is true of their base salary. In fact, the second time I was a visiting professor there, I arrived to teach the Spring semester in the middle of a teacher's strike. They merely wanted their base pay to equal that of the garbage collectors!

But, in truth, the salary situation for most professors, at least for those who conduct active research programs for the Israeli government or NASA, for example, is quite good. They supplement their base salary by drawing directly from the contract work. In the US, most of such contract funds are generally used to support graduate students or buy equipment. The top faculty also supplement their salary by consulting, which is encouraged. The TECHNION gets much support directly from the government. Another nice (?) feature is that the faculty can 'borrow" heavily on their anticipated retirement funds, with no great pressure to pay back.

The Aerospace Engineering faculty, headed by a Dean, who rotates out of the job on approximately 3 year centers, almost universally consists of what I call "Battleship Admirals" – persons who do not recognize that there is a viable regime above the earth's sensible atmosphere. Of course this situation is reflected in aerospace departments throughout the world. Only retirement of the 'old timers' will permit a more benign attitude towards astronautics. The aeronautical buffs hang on even though the airplane business is on the decline. In fact, the Technion's Aero Department was one of the last in the modern world to convert its name from "Aeronautical Engineering" to "Aerospace Engineering" in recognition that the Space Age was probably here to stay. The catch is that the Faculty did not change and, in many cases, remains blissfully unaware that such subjects as flow around an airfoil are of absolutely no interest to a space aficionado.

Dr. Tancum Weller, who was Dean of the Aerospace Faculty in 1999, <u>was</u> interested in space when I first arrived. He is a structures expert and was leading a group of his students in an international design competition. It was sponsored by the NASA/Jet Propulsion Lab of Pasadena. The idea was investigate the possibility of erecting a huge solar sail in outer space which, utilizing the reflection and/or absorption of the high energy particles which are emitted from the Sun to provide very low thrust propulsion, would facilitate travel to nearby planets. The problem was how to erect the sail. One obvious solution was to use inflatable tubes to give shape to the very lightweight sail material. However, the fear of deflation due to particle puncture was a real one in this approach.

Years before, when I was working with the Atomic Energy Commission's lab at Iowa State University, the alloy of Nickel-Titanium was brought to my attention. In wire form, NiTi (pronounced nye-tye) has some unusual properties. It is very ductile, like copper wire. If it is heated to a certain temperature and formed into a particular shape, like a circle, say, and then allowed to cool, it will return to that shape when heated again to the critical temperature. This is true no matter what is done to it while cool; like packing it into a small space by warping and bending it, say. Just heat it, even if it is in a ball, and it will revert back to the set shape, where it will stay when the heat source is removed or turned off. I suggested to the group that they propose NiTi to erect the sail, using electrical energy to heat the wire. They liked the suggestion and proposed it. They were one of the winners.

Throughout the TECHNION, there is a strong belief that it is the fundamentals that must be taught, and that it is industry's role to teach their graduates the applications. This is a highly elitist attitude, which MIT, for example, long ago abandoned. It is sustained by the fact that the faculty has had very little industrial experience and their school philosophy is that every student should go for the PhD. They have a fear of becoming a 'trade' school. But, methinks their pendulum has swung too far. Since I teach only applications-type courses, I was like a fish out of water. Some years ago, because of my trade school attitude, a friend said, "Brodsky, you have no clue of what constitutes an education". Sure, the fundamentals are great, but to see them in action is greater! Maybe I'm wrong, but I did win two national 'outstanding teacher' awards.

Anyway, it was this purist attitude that made the Faculty look down their noses on my courses. Unbeknownst to me, they had reduced the credit for my three-hour-a-week course in Spacecraft Design from the normal 3 credits to 2½, without telling me. That was why so many dropped out after I told them about the homework and exams they should expect. They figured they would have to put out a non-proportional amount of effort for the credit they would receive. Nevertheless, many students stuck by me, and one later became my protégé at USC, where she earned her PhD.

I reprised as Visiting Professor for the Spring 1994 semester, this time teaching my "*Principals and Techniques of Remote Sensing*" course. The second time around, the 'admirals' did me in even worse, again without letting me know before it was too late. This time around, I taught my Remote Sensing class; a course which has no natural departmental home. It can be taught by Electrical Engineering, Physics, or Aerospace - each having a legitimate claim. The Aero Faculty at the TECHNION didn't know what to make of it, so they decided to give it 2 credits. I screamed like a wounded eagle and wrote an angry appeal, which was eventually denied. I felt sorry for the 20-odd students who decided not to take the course that they really wanted. I gave no homework assignments to the 7 students who stayed, going over the answers in class instead. For the final exam, I gave only one question: "Who is buried in Grant's Tomb?" I gave everyone an "A" even though all could not come up with the right answer.

Chapter 6

SNAPSHOTS FROM THE TURN OF THE CENTURY

In 2008, I became a 60 year member of my technical society, the AIAA (American Institute of Aeronautics and Astronautics). I had joined the precursor organization, the IAS (Institute of the Aeronautical Sciences) as a graduate student at NYU in 1948. Some years later, along with my career change into Space, I joined the ARS (American Rocket Society). In the 60's, the two organizations merged to form the AIAA. In 1987, I was selected a Fellow of the Society - it's highest regular rank.

The same year, '86-'87, I became Chairman of the Los Angeles Section, the largest in the country. I have served on many of the society's Technical Committees and its Educational Activities Committee; written and delivered many technical papers, written technical book reviews, formed new Sections, served on the L.A. Section's Executive Council consecutively for over 20 years, and have done various other jobs in its support. Today, I am active in the 'Aerospace Alumni' group of old timers. I have always enjoyed the interaction with people in the business.

Even at the end of 2011 approaching age 87, I remained active with the LA Section and seldom missed an Executive Committee meeting or a monthly dinner meeting. At the latter, I have often taken notes and written a review of the event. This chapter is devoted to reprinting some of the reviews that appeared in the Section's Newsletter. They were selected because they give a good indication of what was happening in the aerospace industry at the turn of the century. I always gave the Speakers a look at the review so that they could make alterations. Most of them ran them through their Public Relations Office so that all would be PC - Politically Correct.

WE ARE ECLIPSED!

Michael S. Kelly, CEO of Kelly Space and Technology, Inc. (KST) of San Bernadino, presented his company's entry into the vigorously competed on-going launch vehicle system sweepstakes at an October 15, 1997 dinner meeting of the LA Section at the TRW forum. His patented concept is named the '*Eclipse*" system, and it is presently being done under the aegis of Motorola as a part of the Iridium constellation of communication satellites. KST will receive around 100 million dollars for the successful delivery of ten Iridium constellation replacement (i.e. 2nd generation) satellites into LEO orbit.

Kelly described the status of his company's program: As presently envisioned, the *Eclipse* system consists of a towed (by a lightly loaded modified 747 tug aircraft), manned first stage vehicle which achieves Mach 8-9 at 400,000 feet following 747 disengagement at 20-30,000 feet and rocket boost, by either liquid or, alternatively, solid rocket propulsion. The Iridium payload is carried in an upper stage booster, disgorged - clam shell fashion - from the nose of the Eclipse vehicle. The Eclipse is then piloted back to the take-off field, with turbojet assistance, following its return to the sensible atmosphere. The original version of the upper stage will be expendable. Later variations may be winged versions which will automatically re-enter and be recovered, either by horizontal landing or by parachute, thus making the entire launch system reusable. During development, smaller demonstration versions of the Eclipse will be built and tested. Construction and detailed design of the Eclipses will be subcontracted to a Santa Barbara firm.

KST has already demonstrated the tandem take-off system, wherein the 747 will be airborne while the first stage is still rolling down the runway at the end of a 5-6,000 foot tether tow line. They accomplished this by using two light planes of approximately equal size. They have also demonstrated successful control characteristics by flying an F106 at various locations in the wake of a C141 cargo plane. Kelly said he anticipated no difficulty in finding qualified pilots to man the Eclipse vehicles. All present admired his confidence, guts and ability to attract venture capital.

(Note: As of May 2004, this program was cancelled; as the Iridium system went into financial decline)

RUSSIAN MOONY TUNES

The January 28, 1998 monthly meeting at the TRW FORUM was an auspicious event, marked by a "Distinguished Lecturer" along with an equally distinguished AIAA member guest list. An audience of 85 attended the talk presented by AIAA Executive Director Emeritus (from 1964-1988), Jim Harford. His entertaining and historic talk discussed *"How the Russians tried to beat us to the Moon"* resulted from research he accomplished over a six year period, during which he made 15 trips to Russia and interviewed close to 50 Soviet engineers, scientists, family members, and politicians involved in the USSR's early space efforts. The result is a newly published book entitled *"KOROLEV, How One Man Masterminded the Russian Drive to Beat America to the Moon"*. His talk consisted of interesting illustrated excerpts from the book highlighted by personal observations and anecdotal background stories. John Wiley & Sons is the publisher.

Sergei Korolev was both the anonymous (at that time "the Kremlin feared he might be 'done in' by the CIA") and legendary "Chief Designer" of the USSR's space endeavors in the late 50's and early 60's. He and his group were responsible for Sputnik, and the first dog, man, woman, group of men in space; and later the first spacecraft to impact the moon and Venus. Spurred on by these successes, he then tried to mount an effort to beat us to the moon. Harford explained that the effort was too little and too late. They didn't have the propulsion capability that was anywhere near the F-1 engine's thrust. Korolev's recognition by the public did not come until after his untimely death. Now, he has a city named after him, as well as a handsome, imposing statue. Thanks to Jim Harford, his story is now known in the English speaking world.

Appropriately, Harford was introduced by a space-age giant, Honorary Fellow Dr Bill Pickering, who - almost forty years ago to the day of our meeting - presided over the USA's first successful space flight and was AIAA's first President over AIAA's first Director- Jim Harford. Both had held similar positions in the American Rocket Society, which merged with the Institute of the Aeronautical Sciences to form AIAA.

Other high profile attendees were Honorary Fellow Lee Atwood, former head of North American Aviation, Fellow member Dr. Abe Zarem, who founded the pioneering Electro-Optical Inc. company, and Dr Louis Friedman, Director of the Planetary Society. All were welcomed by Section Chairman, Bob Sackheim, while Bob Brodsky acted as emcee. Following the talk, Harford autographed 28 copies of his book.

A TRW LIGHT & ZAP SHOW

The February 25, 1998 monthly meeting featured Dr Tom Romesser of TRW speaking on "High Energy Laser Technology". About 42 members and guests attended the TRW Forum meeting and dined sumptuously on chicken parmesan and high tech gimmickry and planning. Tom is Vice President and Deputy General Manager for Laser Programs in the Space & Laser Programs Division, having been appointed to this position in December 1997, following a succession of increasingly responsible jobs in this field since joining TRW in 1975.

Tom told the audience that the chief industrial applications of high powered lasers (HEL) are for cutting, welding and, in a defensive military stance, for weakening, heating, and ultimately destroying a target. At present, there are three ongoing programs in which TRW is participating as the HEL contractor: a tactical (i.e., ground or sea based) program (THEL); an airborne program; and a spaced based program. The high powered lasers (> 100 kw) have been under development for many years and have their basis in a continuing research program, called Alpha. Hydrogen- and Deuterium-Fluoride (ugh!) provide the working fluids for lasing. The latter provides energy at 3.8 microns enabling passage through the atmosphere.

THEL, the most advanced program, is sponsored by the Army and Israel and will come into a test stage this summer against artillery rockets. Other design targets are UAVs, cruise missiles, tactical ballistic missiles, and aircraft. Not only must the target be hit, but it must be hit in a vulnerable spot or band (if it is spinning) to preclude fragments from hitting the intended target. Tactical stations will be established along the incoming flight path between known adversaries and their target, as far towards the adversary launcher as possible.

As in all the TRW programs, Lockheed/Martin is developing the fire control systems. These two companies, teamed with Boeing as prime, are developing the 747 air based system under Air Force contract. The program is aiming for a demonstration in 2002 and ultimately a dispersed fleet of 7 aircraft. Here, the targets are theater ballistic missiles - hopefully during their boost phase. COIL (chemical oxygen-iodine) lasers are employed with sufficient fuel onboard to allow maybe 40 individual attacks. Adaptive optics, featuring deformable mirrors are needed to focus the beams compensating for atmospheric caused optical aberrations.

The USAF space-based program is also aimed at boost phase intercept with demonstrations hoped for in the 2005-08 period. Lockheed / Martin is prime, with TRW laser system support. The ALPHA program HF laser (i.e., shortest wave length to maximize kill probability) will be employed. The LAMPS (Large Aperture Mirror demonstrator Program) features a 4 meter controlled figure mirror system. The critical spacecraft stabilization and pointing tasks are being done under aegis of the Albuquerque Air Force, and breadboarding is already under way.

Surely, this is the shape of things to come. Already, a new business area dealing in counter -and counter-counter measures is under consideration. Stand By!

THE AEROSPACE FORCE IS WITH US

General (ret.) Bernard Randolph enlightened and delighted 45 Section members on "Future Air Force Roles and Missions in Space" at a TRW Forum dinner meeting on March 19, 1998. The General, now a consultant, recently retired from a vice presidency at TRW, following a 35 year career in the Air Force which culminated as Commander of the Systems Command. The "delight" part of the program came at the end, when the audience wouldn't stop asking, badgering, and commenting. Great Fun!

Gen. Randolph said that the Aerospace Force - a new name they are trying to promote - is seeking to do things in a radical new way since their entire budget, now under discussion in Congress, is the same as his Systems Command budget when he retired over 10 years ago. Acquiring and managing information is their key to success, and they are trying to trade off operational funds (by closing down unnecessary bases) for R&D funds to achieve this goal. In conjunction, they are attempting a revolution in the way they manage their business affairs by streamlining acquisition and procurement practices and taking a "system of systems" approach. This latter approach is trying to eliminate inter- and intra-service duplications, as well as using the best systems that each service and the commercial world has to offer. He noted that the latter segment is becoming huge- far overshadowing the former military dominance of space, thus offering new opportunities for the services and relieving them of some R&D efforts.

In an admittedly controversial part of his talk, he stated that control of space and force projection from space would be required. This would mean the almost instant ability to have global awareness and presence - engendered by the ability to obtain information from many sources on a real time basis. Analytical abilities and rapid decision making would be part and parcel of this package. Such on - demand real time awareness would be obtained by space based radars and other hyper imaging systems. The knowledge would be backed up by the ability to deliver munitions from space.

Randolph further stated that the control of space would depend on surveillance, force projection, and counterforce operations. He also cited a new field - "Information Warfare" - wherein space spin doctors would be turned loose to confound potential enemies. He noted that a key to these goals was a stable of reliable, low cost launching systems, including SSTOs (single stage to orbit) and other reusable systems. The areas where civilian applications were not footing the bill will have to be borne by the Services. The Aerospace Force is again looking at on-orbit maintenance by robotic means, with SSTO vehicles playing a hand in this for reasons of quick response.

The talk was interesting and controversial. We shall now see what the new millenium brings!

SOJOURNER BEWARE !

"Thar's danger in them thar orbits !", so said our May 20, 1998 meeting dinner speaker, Dr. Val Chobotov, longtime manager of Aerospace Corp.s' *Space Hazards Section*. Or, less alarming, trouble to be expected in the not too distant future if we don't take already prescribed mitigation steps. The meeting took place at the TRW Forum and was attended by about 45 fans of impending disaster.

Chobotov concluded that the present day threat of natural hazards and man-made orbital debris is not large, but does threaten to increase greatly in the near future as the new constellations of satellites become reality. And, he said, when it does become a threat, it will be a big one. It is much cheaper to take preventative action now, rather than wait until such action becomes necessary. In this regard, some areas of concern need timely resolution: Better numerics on both the LEO and the GEO threat; The anticipated impact of the new multi-satellite systems; Collision avoidance and protection methodology; and Reentry debris break-up design considerations.

The speaker first defined the natural environment (transient micrometeoroids, and storms and streams traveling at around 20 km/s) and the man-made environment characterized by delta V's of around 10 km/sec max. This latter debris, which remains in earth orbit, is arbitrarily divided into "large" (> 10cm) and "small"(<10 cm) particles. There is thought to be about a total of 1000 kg of material less than I cm and about 300 kg less than 0.1 cm presently buzzing merrily around the earth. As of the first of the year, around 8000 objects greater than 10 cm have been noted by a variety of space and earthbound systems, including Geodds and other optical tracking systems and radars. Half of these are fragmentation debris. The debris density (particles /cubic meter) has twin peaks at 1000 and 1500 km altitude, and mostly stems from polar orbiting systems. A lot of data came from the recovered LDEF experiment, which suffered over 34000 impacts during its 6 years in LEO.

Nevertheless, Chobotov stated that the collision probability with large debris is small. The hazard assessment has been mathematically modeled and formulated. For DOD assets, the collision probability with large objects is 1 in 10,000 years, and for objects less than 1 cm, 1 in 500 years. For the ISS space station, the probabilities are, for < 1 cm, 1 in 95 years, and for < 10 cm, 1 in 1000 years. The ISS is being designed for survivability, and uses shielding and monitoring as defense mechanisms. Both NASA and Aerospace Corp. have developed severe environment, density and risk program models of the space debris environment. Moreover, several experimental facilities, featuring light gas guns and other accelerators, are performing simulations. Sandia has a 12.2 km/s facility; the University of Alabama a 16 km/s one, along with NASA ranges. Alas, some other-worldly debris travels as high as 70 km/s!

The problem is an international one and is recognized as such, with suitable UN Committee oversight. In the United States in 1995, the National Research Council made a major study and provided several recommendations, which have been followed through on. The USAF Phillips Lab is the leader in a program with three main thrusts: Measuring, Modeling, and Mitigation.

The mitigation effort is coordinated and has produced both NASA and DOD guidelines for design. The guidelines cover minimizing the future threat, disposing of debris, control of debris production, and proper selection of flight profiles. In January, 1998, a draft *"US Government Debris Mitigation Standard Practices"* document was floated and is now in the review process prior to becoming law.

Not only did the meeting provide great divertissement, but the menu also provided a mystery. Although good, I have absolutely no idea of what the two (your choice, or if you were brave, both) entrees were. Perhaps I'll offer a prize at the Section picnic in July for the one who can come closest to the formulation!

WHAT A GREAT CLAMBAKE IT WAS !

The Section's 1997-98 season meeting agenda came to a golden close on June 17, 1998 at the "LOW COST LAUNCHERS FOR SMALLER PAYLOADS" dinner meeting at the TRW Forum. In fact, at around a head count of 110, it was the best attended Section meeting in this old timer's memory, which dates back to 1958! Five knowledgeable and erudite speakers represented the launch vehicle aspirations of the following companies: Orbital Sciences Corp. of Dulles, VA; Rocket Development Co. of Los Alamitos, CA; Microcosm, Inc. of Torrance, CA; Rotary Rocket Co. of Redwood City, CA; and Kistler Aerospace Corp. of Kirkland, WA. Inexplicably missing, due to a last minute renege, was Lockheed/Martin of Denver, CO. (Athena family of vehicles). Earlier this year, we had heard from Kelly Aerospace's entry in this new space race. As you will see below, the panel included people who have lived in Tokyo, Moscow, and in Space. I will now try to briefly summarize each of the approximately 15 minute presentations:

ED MORRIS, OSC

Ed is the OSC Director of New Initiatives and a USAF veteran of Titan, Delta, and Atlas launch operations. He described Orbital's three active launch system programs: The Pegasus, the Taurus family, and the X-34 demonstrator which they already offer or are developing. The Pegasus is a conventionally stacked three solid motor staged L-1011 air-launched vehicle whose first stage is winged. It has been launching the OSC ORBCOMM satellites, among others. There are three tail sitting Taurus versions planned. The first, which has already flown, adds a new first stage to the un-winged Pegasus and obtains about 3 times the Pegasus payload capability. Two later hybrid versions with more capability have one liquid upper stage. The X-34 RLV is a reusable, rocket propelled, unmanned, winged Mach 8 demonstrator which will reach altitudes of around 250,000 feet. OSC reduces costs by simplified, unsophisticated launch platforms and offers customers a complete turn-key operation.

TK MATTINGLY, RDC

TK is President of Rocket Development and a former Apollo and STS astronaut- command pilot, as well as a retired Rear Admiral. He carefully explained the difficulty of achieving low cost in what he characterized as a "thin" market. He said that low firing rates caused fixed expenses to dominate over unit costs. Thus, RDC is aiming to capture the most lucrative commercial markets where many launches will be needed. He believed that the correct solution to keeping down costs is to start from scratch and to amortize development costs, and to reduce labor costs by stressing cost effective approaches. RDC is developing three versions of the Intrepid ELV launch family destined for Polar LEO deliveries, with the first demo flight planned in the year 2000. All are 2-stage stacked Lox-Hydrogen powered vehicles, capable of delivering 2,100, 7,500, and 11,000 pounds to orbit.

Dr. BOB CONGER, MICROCOSM

Bob is the Vice-President of Microcosm and is in charge of their Scorpius ELV program, and is a former President of Logicon - Intercomp Inc. and a former Air Force Special Weapons Lab employee. The all-liquid, Lox-Kerosene, pressure fed, Scorpius six vehicle family, now under development with Air Force assistance, includes a sounding rocket, a sub-orbital vehicle, and 4 orbital launchers including a Minilift vehicle; the Liberty Light Lifter; and Exodus Medium and Exodus Heavy (15,000 pounds to LEO for $15 million is the goal) vehicles. The orbiting vehicles have a common core and up to 6 attachable pods of equal size. There are three or four sequenced stagings. At present, control is by LITVC and/or thrust modulation of the throttle-able engines, although gimballing is also being investigated. The engines are supplied by Rocketdyne and Aerojet, as well as by an original 5K# Microcosm design. Testing of the larger (20, 80, and 320 Klbs) engined designs will take place at White Sands.

GEOFFREY HUGHES, RRC

Geoff is the Director of Business Development for Rotary Rocket, and had the foresight to write what is probably the first academic thesis on how to achieve low cost launching. He was educated in England, has taught in Canada, and was manager of strategic marketing and international operations of a leading Tokyo electronics and telecommunications firm. RRC's ROTON is a reusable-piloted reentry SSTO Lox-Kerosene spacecraft, aiming for 3rd quarter, year 2000 initial flight. It is a tail sitter for both take-off and landing. Its development, like most of the others, is privately funded and they are seeking commercial-only customers. Its low cost potential lies with its total reusability and its capabilities are to lift 7000# to 200 km at up to 50 degrees inclination. The ROTON has two unique features:

(1) Its engine configuration is of the aerospike variety with thrust chambers distributed around the periphery. The engine is mounted on a bearing and rotates at 720 rpm. The propellants are introduced down the middle of the bearing, where centripetal force increases their pressurant-created pressure to a high value, acting like a centripetal pump, before they enter the thrust chambers. Rotation is caused by canting the thrusters, and body counter-rotation is achieved by exhaust gases;

(2) Following reentry with proper alignment assured by fins, large autogyro blades are deployed and soft land the vehicle at the launch site. Many contractors are involved, and some subsystems are taken directly from other vehicles. The key body parts, mainly graphite reinforced fiber materials, are being made by Scaled Composites of Mojave, CA. RRC is additionally offering on-orbit operations and space salvage.

DEBRA FACKTOR LEPORE, KISTLER

Debra is KAC Manager of Payload Systems and acts as technical liaison between payload system requirements and launch vehicle development. She was previously ANS-

ER's Chief of Moscow Operations. Kistler will offer two (actually they are building 5), 9-day turnaround, fully recoverable/reusable vehicles for commercial use. An impressive board of directors is coordinating the work of their major subcontractors: Lockheed/Martin, Aerojet, Draper Labs, Allied Signal, Irvin Airchute, and Northhrop-Grumman. The stacked vehicles are designed to carry around 5000# to 400 km, with all three stages recoverable at the launch site, using a combination of multiple parachute clusters and air bags. Aerojet is reworking the pressure fed Russian- built Lox-Kerosene engines; the GPS navigation/control system is off-the-shelf by Honeywell; both being factors helping to keep down costs, estimated to be 17 million per launch. Other cost reducing factors are the use of aluminum tankage and modular construction.

At the end of the formal presentations, a vigorous Q&A exchange took place, followed by RRC and KAC tape viewings. The moderator (little old me) expressed his opinion that as good as all candidates sounded, it was sad to think that there may only be one or, at best, two winners. I also thought, with a sigh, that the day of the "Big Dumb Booster" (i.e., pressure-fed LITVC-controlled Lox-Hydrocarbon engines with lightweight construction) was finally arriving in a fair competitive framework. But, will it win? Sic Transit Gloria!

(Note: As of Dec. 2009, only the Microcosm and Orbital Sciences efforts remain viable, but new entrants – SpaceX and Rutan/Branson are now, 2009, major players)

QUO VADIS LOGISTICS ?

Man and boy, in my 46 active years in the biz, I hardly gave a second thought to LOGISTICS, tacitly relegating this area

of systems engineering to the same dungeon I had previously and, it now appears, just as mistakenly, placed the newly emerging fields of reliability and value engineering as they reared their ugly heads in the 60's. Will I never learn?

My head was properly shrinked (shrunk?) on Dec. 16, 1998 at the TRW forum at our pre-holiday monthly dinner meeting. About 13 others, most of them being aficionados of the sport and thus knowledgeable on what logistics entailed, heard Gerry Facon speak on "Logistics in Transition". The speaker works at Lockheed Martin Technical Operations and is a veteran of 35+ years of experience in Logistics Engineering, Systems Engineering, and Information Technology Management. He is also the current secretary and Chair-elect of the AIAA Space Logistics Technical Committee (SLTC).

Gerry spoke briefly on SLTC activities and focused on the need for greater emphasis being put on up-front supportability design influence, since Logistics is sometimes treated as an afterthought (but actually may account for over 60% of the total cost of lifetime ownership). In some detail, he discussed the several objectives of the Committee, including their major concern to assure that their inputs are included in the systems engineering thinking at the very beginning of a program. Before the recent implementation of new government acquisition processes, a separate Logistic Plan (Mil. Standard MIL-PRF- 49506) normally was part and parcel of all RFPs. Now, such plans are melded into the family of concerns that Systems Engineering encompasses, and the Committee fears that support for its work is waning and its membership is falling.

A major thrust of the discussion was the fact that the Logistics disciplines, as we knew them in the past, are being absorbed by other disciplines, such as Systems Engineering, or into Integrated Product Teams, which have too little appreciation of the value of Logistics. This situation is especially true in the area of space logistics, which largely grew out of the laboratory environment, whose practitioners had no fundamental knowledge of what Logistics even meant - except when they needed a spare part.

Since my knowledge of what roles logisticians play was mostly intuitive, I asked lots of questions - and a better picture of the field emerged: A short but good all encompassing description of space logistics can be found in Chapter 19 of "*Space Mission Analysis and Design*". While many universities teach the individual components of logistics, only a hand full of schools, such as Embry - Riddle University, address the full spectrum of Integrated Logistics Support by providing a course in Space Logistics. The best complete text on the subject is by Blanchard and is entitled "*Logistics Engineering and Management*", which describes the ten components of logistics: Design Influence (the focus of the speaker's discussion); Logistics Support Analysis, which is the logistics process driver; Training; Technical Data; Supply Support (those spares guys); Manpower/Personnel Requirements; Packaging, Handling, Storage & Transportability; Facilities; Computer Resources Support; and Support Equipment.

The speaker said that the B-2 (airplane) program represented the epitome of correct application of logistics engineering in the aerospace business, and that the space sector was just learning. He discussed the "rich implementation" of Integrated Logistics Support (ILS) in the B-2 Program, where long-term supportability objectives were not sacrificed for short-term cost or schedule considerations. As a result, the B2 is a highly reliable and maintainable vehicle, and is already seeing the benefits of a well thought-out Pre-Planned Product Improvement Program. Similarly, the B2 ILS Program incorporated Reliability and Maintainability Engineering into a true

Integrated Product Team environment. The result was the B2's award winning Logistics Support Management Information System program, which exceeded CALS (Continuous Acquisition & Life Cycle Support, a.k.a. Computer-Aided Logistics Support) requirements before anyone even heard about CALS, now more commonly known as Electronic Commerce.

Logistics is changing and an inevitable marriage between "Supportability Engineering" (which many look at as the umbrella over all RMA - Reliability, Maintainability & Availability- and ILS functions) and Systems Engineering, which focuses on requirements management and risk management. The speaker allowed that this does seem like a good marriage, and could help slow the "Development Death Spiral" where we keep building more costly and more complex products, which are more costly to support, leaving less and less dollars available for new development and modernization programs.

But, since logistics is such an integral part of the Life Cycle Cost equation, logisticians believe that their discipline will survive intact and perhaps, as management gains more knowledge of their worth, prosper. In my opinion, to make this happen, they must give the concept of supportability more media hype, and foster more educational and paper presentation opportunities. Should anyone be interested in joining the Space Logistics Technical Committee, or simply need more information on the topics discussed, Gerry can be contacted at (310) 727-1088 or Email <gerald.a.facon@lmco.com>.

On a lighter note, the carte consisted of excellent fried chicken, with baked beans, cole slaw, and roll and butter. Where else can you get all this for a mere ten dollars?

WHAT DONE IN TWA 800 ?

I don't know what happened to TWA 800. And, Aeronautics Professor Joe Shepherd of Cal Tech doesn't know what ignited a lethal mixture of air and fuel vapor in the center wing tank, but given that ignition occurred, he has learned a lot about the disaster that followed. Only my friend Bill Dixon, who advises not to discount a meteorite collision, and Pierre Salinger, who early-on reported that it was an awry Navy missile what did that good ship in, know for almost sure.

On February 17, 1999, Shepherd talked to 48 attendees at the LA Section's monthly meeting at the TRW forum. His topic was "Fuel Flammability and the Crash of Flight 800". He is currently leader of the Explosion Dynamics Laboratory, which does pioneering research in combustion, explosions, and shock waves. Shepherd has built his career on experiments, analysis, and computational and simulation methods on explosion phenomena. His doctoral thesis (1980) was on "The Dynamics of Vapor Explosions". He worked for Sandia Labs and taught at RPI before returning to Cal Tech in 1993. Thus, it is not surprising that the National Transportation Safety Board (NTSB) sought out and is funding his services to help solve the mystery surrounding the accident of July 1996. His investigation has been ongoing for two and a half years, and should be completed this year.

The explosion occurred about 20 minutes out of JFK at 13000 feet. Because of its prior flight history, fuel from the center wing tank (CWT) had been mostly consumed during the inbound flight, and refueling of this tank was not needed for the flight continuation. Most of the wreckage was recovered in water that was 100-150 deep. Over 100,000 pieces were recovered in the biggest salvage operation ever, and were reunited in an old Northrop-Grumman hangar for forensic study by teams of experts led by the NTSB, including specialists from Boeing. It became clear that there had been an explosion in the center wing tank, a 5-longeron, 6-wet bay wing carry- through structure. The bay compartments had openings in them to permit fuel to flow through them.

Shepherd worked with the explosion analysis team to determine what range of air / fuel vapor mixture was volatile; to determine the relation between flammability and ignition, and to determine if an explosion in the CWT could cause the dramatic failure. To assist in this, Boeing and the NTSB instrumented similar aircraft with thermocouples to investigate the temperatures at the tank's surface. Using these data, Shepherd's group performed over 60 quarter-scale CWT explosive tests to determine how the flames propagated through the six tank bays to provide information to simulate the structural failure. Other research groups are examining how to simulate the structural failure and flame propagation through the use of full scale tanks.

The investigations showed that the three air cycle machines mounted directly below the CWT produced hot spots with temperatures up to two hundred degrees F. Such temperatures were found to be more than adequate to vaporize enough fuel to produce a flammable mixture. All that was necessary was a low energy spark to start the explosion. The model testing convincingly showed that portion of the carry-through structure would be damaged no matter where the initial ignition took place.

Professor Shepherd said that the NTSB is considering all possible ignition sources, and is particularly wary of wiring associated with the fuel quantity instrumentation system. They have already suggested modification to reduce the heat load received from the hot air handling equipment sited below the CWT. To my mind, the shocking statistics that Shepherd quoted make me think that they are looking for a needle in a haystack- and that the Dixon and Salinger theories have as much credence as anyone's: There have been over 317 million commercial aircraft departures since 1959. There has been one fuselage related loss per 36 million take-offs! All such failures except TWA 800's have been understood and only one, in a 737, has been attributed to CWT failure. Make your own conclusions!

The cuisine this evening was of the Chinese persuasion. It was good, but by the time I got in line (I slaved 'in the office' at collecting the attendee's money) it was a bit tepid.

THE TRUE APOLLO 13 STORY

I must be accident prone! Last month I reviewed the TWA 800 tragedy; herein lies the Apollo 13 "accident" un-bared. Accident in quotes because the speaker's former mentor, Admiral Rickover said, "There are no "accidents"- only "incidents" stemming from poor quality control". In a multi-media blitz, using machine-driven view graphs, audio, and splices from the Tom Hanks Apollo 13 movie, AIAA Distinguished Lecturer Owen Brown drove this point home to about 60 attendees at his March 17, 1999, TRW Forum Dinner Meeting talk: "*Apollo 13: The Rest of the Story*".

The funny thing about Owen, now in the home stretch of completing his PhD requirements in Aero/Astro at Stanford, is that he was in 2nd grade when the Apollo adventure took place. He picked up on it by reading all he could about the near-disaster during a three month cruise while he was a nuclear - trained submarine officer onboard USS Flying Fish in "Rickover's Navy". In particular he cites the NASA accident report and a book, "*Apollo, The Race To The Moon*" as major sources of his later inspiration and determination to really understand the causes of the of the ruination of the Lovell, Mattingly, and Swigert flight. In his analysis, he draws on his observations and lessons learned from the rigid control disciplines exercised by our nuclear navy. His presentation gelled while he worked as a spacecraft engineer at Loral from 1993 to 1998, whence he continued his education, having previously earned an MS degree from Stanford.

It had always been my thought that there had been a detonation/explosion that caused the failure. This turns out to be not quite true. What really happened was a successive failure chain of events that ultimately caused the pressure in one of the main oxygen tanks to rise so high as to cause an overpressure failure. Owen then carefully tracked down the mistakes that were, but shouldn't have been, made. The culprit #2 oxygen tank was a starboard side 2 foot radius double shelled Inconel tank which held the O2 at −300 F. Two electric motor-driven stirring fans were located in the tank, as were fill, drain, and outlet lines, and an emergency vent valve which springs open in the unlikely event of overpressure. The reconstructed history of tank #2 is lurid, and undoubtedly has caused Admiral Rickover to roll over in his grave. It was originally intended to be installed in Apollo 10, but was inadvertently dropped about 2 inches during installation and scrubbed from that mission. A routine visual inspection found no dents or other obvious scars and so it was reassigned to Apollo 13. Now, the plot thickens! :

Three weeks before A-13 launch, during a dress rehearsal, the total oxygen used was about half of the total capacity, but it was found that very little had come out of tank #2. The thinking now is that maybe a connector was loosened due to the drop. Following the test, the oxygen was drained from the tank under conditions that the overpressure valve, rated for 85 degrees max., saw more than this and welded itself closed. The switch was designed for 28 volts and no one paid attention when Apollo voltage was raised to 65 volts. A later test showed that 48 volts were enough to fuse it closed. So, Apollo 13 was launched with a loosened fill line; a no good switch; and exposed fan wiring. The latter probably occurred during the dress rehearsal unloading overheat.

So, when Rusty Swigert threw the tank #2 fan on switch for the first time, the electricity started burning the insulation and the temperature started rising. The emergency switch was already inoperative, and it is speculated that the temperature may have reached 1000 degrees before the blow!

Owen finished up by summarizing the lessons learned: Not only must a strict Quality Control regimen be enforced, but a quality Attitude is needed by all involved! On the other hand, Mrs. Lincoln, the menu of fried chicken and beans stood muster - not wonderful, but good and filling.

COME FLY WITH ME

Patrick Carey, production test pilot for Gulfstream Aerospace Corp. of Long Beach, FAA Designated Pilot Examiner, and Chief Instructor Pilot of Security Aviation flight school located at Northrop Field was the featured speaker at the October 20, 1999 dinner meeting at the Northrop Rec. Facility. His talk was scheduled to be "Obtaining Your Private Pilot's License". We supposedly designed the meeting to attract young members who presumably would want to know what it takes to get into recreational flying. Oh, Boy - were we wrong!

Instead, about 10 'Old F(ellows?)', all presumably flying buffs, showed up. So, agile on his feet, Pat sort of changed his theme to the possibly more delightful, 'My Life and Hard Times in Aviation', which more or less described how he got to be where he is happily now, without becoming an airline pilot. But, before I tell you his story, let me give you his original talk's bottom line: These days it takes about $4000 to get your Private license, and you can't try soloing until 15 hours of dual time. Contrast this with my story from 1944, when I soloed after 8 hours of dual. Then, you could get dual instruction on a Piper Cub for $3 an hour, and solo for $2 an hour, and spend ~$300 for your license!

Pat, a local boy, learned to fly in '62-'63, and got his Commercial ticket in '65. He lost his student deferment from the 'Nam fiasco and ended up in the infantry. Fortunately, he was assigned to the Army Flight School in Texas, and after 40 weeks training was a genuine helicopter pilot Warrant Officer. Off to war, he put in 930 hours flying a Cobra gun ship, and was only shot down once. He came back and spent 3 years as a primary helicopter instructor, got promoted to Captain in the Artillery, got a BA and a Master's at USC on his own, finally got a fixed wing Army job, from which he retired.

In 1987, he became a corporate pilot at Northrop and eventually was their Chief Pilot, towards the end of his 12 year association. During this time, he flew many strange planes and flew several 'strange' missions. In 1994-5, he joined his fellow pilots to turn the Northrop Flight Department into a separate commercial operation, which after a rough start, is now very successful as Northrop-Grumman Aviation. They run 2 King Airs, a Twin Otter cargo plane, a LearJet, and a Citation 10 for Northrop's CEO.

In July this year, an old friend, using money and new challenges as coercion, got him to join Gulfstream, who produce the Cadillacs of bizjets, the G-4, and the top-of-the-line G-5. These are manufactured in Savannah and flown to Long Beach for final painting, outfitting, and check out before delivery. Pat and two other pilots do the flying, sometimes carrying as many as 10 technicians, each checking the performance of their particular subsystems. The G-5 is very advanced and has all the bells and whistles: a GPS-based navigation system, a Heads-Up Display for IR night landing assistance, and a 'talking' system which reports out altitude and height above runway on landing, for example.

On the weekends, Pat trains flight instructors at Security Aviation, and is often called on to perform FAA sanctioned Pilot Examination flights, where he grades the performance of private and commercial pilots. He keeps busy, and is a very happy man- pleased with what he has accomplished in a business that he loves. He says that there is a critical shortage of commercial pilots now, and he dearly needs Flight Instructors for his school. He will provide information and training (310) 676 2206.

All the 'Old F(ellows?)' present had their tongues out. What a great thing to do but, alas, Life has passed us by! Oh, by the way, dinner was fried chicken, beans, and cole slaw. Nourishing, but not gourmet

THEY'RE BACK !!!!

An ongoing program that will strike fear and loathing into the hearts of the MUFON / UFON aficionados was described in fascinating detail at the January 17, 2001 Section Dinner Meeting at the TRW Forum. For Gordon Ow, CEO and proprietor of GO AIRCRAFT, LTD. (GOAL) is, as we speak, developing "The High Speed Vertical Takeoff and Landing (HSSTOL) Aircraft" and it's a dead ringer to a traditional flying saucer but for the appendage of two underslung jet engines that are attached to the non-rotating central core.

Mr. Ow described the program, whose first two phases have been funded by DARPA and supported by NASA, to an enthusiastic audience of about 35 who had been very satisfactorily wined and dined with Chinese cuisine. As suggested by the title, the UFO-like vehicle rises vertically driven by an enclosed fan system which rotates on an air bearing which joins the central core with the complete fan-encompassing periphery. The thin toroidal outer saucer section is spun, turbine-like, by air which is diverted via ducts from the engine. Once up to speed, the momentum engendered by the spinning outer section, has converted the vehicle into a large gyroscope. When the fan blades are rotated into a lifting configuration, the outer section takes air in through the upper surface and accelerates it downward via fan action to provide vertical lift; while the gyroscopic effect automatically provides stability in pitch and roll.

The central core features the usual topside glass cockpit-like bulge and, in one troop- carrier version, shows a passenger section behind the pilots' control area. Transition to horizontal flight consists of bypassing more and more propulsive air to accelerate the vehicle forward while reducing the fan rpm and relying more and more on the disc body to provide aerodynamic lift, as the forward speed increases. At the completion of transition, the fan blades are again rotated such that the saucer provides a smooth undersize wing surface. Once in forward flight, directional control is enabled by the usual 'bank and turn' method, achieving the bank by properly precessing the gyro-plane by air jet or aerodynamic controls affixed to the central core. Engine-out fail-safe is achieved by turning the craft into a conventional glider, completely stopping the rotation of the outer section and lowering the rudimentary skid-type landing gear which is normally used for vertical landings.

The Phase I program started in May 1997, and used a 15 foot diameter straw man model, with maximum fan rotation speed of 750 rpm. Basic principles were successfully demonstrated, including wind tunnel tests. Performance estimates showed a much higher forward speed and load-carrying potential than could be obtained by a comparable helicopter, ill-starred Osprey or heavy Harrier-type vehicle. The Phase II program is close to completion. In it, the detailed design of a 45 foot diameter vehicle was studied. Aluminum and fiberglass are main construction materials. Because of the larger diameter, considerably lower fan rpm speeds are required. Under a separate program, the Draper Labs perfected the gyroscopic-based flight stability and control system. Both air jets and aerodynamic supplementary controls are employed. Rotation of the central core caused by air bearing drag is motivated by air jets.

Funding to proceed on a flight demonstrator is being pursued. Ancillary studies indicate that a large diameter lifting fan/gyroscopic principle first stage launch vehicle design may also represent a novel way to go. I was intrigued by the concept, even though the end result is an airplane of less than optimum lift-to-drag ratio. Its compactness, efficiency as a VTOL craft, and versatility for a number of missions, both unmanned and manned, make it an attractive 'what if?' package.

GARDNER GIVES SPACE SKINNY

The Section's Fall 2001-02 season Kick-Off dinner meeting featured Colonel William Gardner, Director of Development planning for the USAF El Segundo-based SPACE AND MISSILES SYSTEMS CENTER (SMC). The Colonel addressed about 40 attendees on *"Future Military Space Programs"* at the TRW Forum on September 19, 2001. Bill heads an organization that performs early planning of such advanced programs as the Space Based Radar (SBR), Global Multi-Mission Service Platform, Advanced Space Lift programs, to name a few of the significant upcoming space assets.

His talk was timely – it called for a new posture for the Air Force in preparing for the unpredictable. Anticipated reorganizations will put space in the forefront of a new defense strategy based on much faster reaction ("sensor to shooter in 1 minute"!) to a range of threats, including the newly added one of Terrorism. Space assets will be closely integrated into total defense strategy. This will be accomplished by the naming of an Undersecretary of Defense for Space; by having the USAF Space Command incorporate developmental planning more closely, by adopting some of NRO's (National Recon. Org.) fast acquisition processes and, indeed by bringing NRO and other intelligence organizations into much closer harmony under the new auspices. More commonality, a la the Multi-Mission Platform, amongst all users – military and commercial - will also be sought.

The present focus areas dealt with advanced war-fighting systems capable of global vigilance. Heading the list were SBR and the visual/IR range hyperspectral sensors. Other important future systems include protection of space assets; space launch on demand; deterrence and power improvements; space transportation R&D; and science and technology and enabling technology breakthroughs.

Finally, Bill had some advice for industry: Keep open lines of communication with the DOD; balance near-term vs. long term investment in technology (ie, the IRAD programs); and retain and grow your technical expertise. A lengthy Q&A session followed.

The meeting was held in the more intimate TRW Exec Dining Room, with dinner sitting in the outside patio – a format improvement but at a $15 cost for members – up from the past $10 dinners. This is a recession?

THE AIR FORCE HAS A FULL PLATE!

The April 24, 2002 Section Dinner Meeting featured a discussion of the USAF's major ongoing space programs. The speaker, a man in the very middle of the action, was William Maikisch, the long-time senior civilian executive officer of the Space and Missile System Center (SMC), at the El Segundo Los Angeles Air Force Base. Bill is heavily involved in all aspects of the daily activities at SMC in carrying out the program of the Air Force Space Command. SMC is responsible for the research, design, development, and acquisition/R&D testing of all satellite, command and control, and launcher assets under USAF direction. His message was loud and clear: In a time of major transition from old systems to new, some radically advanced, the Air Force is facing unprecedented technical and manpower challenges in the face of a tightened budget.

Recent organizational changes have streamlined SMC's ability to act decisively by placing it under the aegis of the Space Command and allowing it to manage both black and white programs. Its budget comprises 90% of the Air Force's space budget. All major programs are now run out of the LAAFB, which has about 2500 employees. Other significant SMC bases (and their population) are Hill AFB (ICBM operations) (878), Kirtland (205), Vandenburg (98) and the Cape (62). Even so, Bill pointed out the severe problem in training and replacing their veteran program management officers, who seldom can stay on a program for more than 4 years.

An even more dicey problem is taking place right now and will continue for the next 2-4 years: The phasing out of 'old' systems and the bringing on line of the new generation – a series of very sophisticated spacecraft, launch vehicles, and support services. Bill cited seven areas where such action was taking place:

• COMMUNICATION AND CONTROL – out with Milstar and DSCS, in with wideband systems
• NAVIGATION – new anti-jamming version and specially coded GPS under study, including consideration of GEO operation
• WEATHER – replacement of DMSP with NPOESS (National Polar Orbiting Environmental Satellite System; with other government agencies participating

- **SURVEILLANCE AND THREAT WARNING** – out with DSP and in with the very difficult SBIRS Hi and Lo constellations
- **LAUNCH VEHICLES** – using up old Titan's, Delta's, and Atlas' and bringing on board the new EELVs: the Delta 4 and Atlas 5 series
- **LAUNCH RANGES** – modernizing and re-engineering Vandy and the Cape
- **SATELLITE CONTROL NETWORKS** – modernizing and integrating

Maikisch also pointed out that there were new tactical issues to deal with. They are negotiating the erection of a brand new facility across the street on Aviation Blvd. They are managing the reorganization of space launch and range management under SMC's command. But most important and vital is the proper integration of all these systems (called 'System of Systems') including the needs and assets of the aircraft and other non-space parts of the Air Force and the other services. And, added to this are international considerations with our European and Japanese friends. It's enough to make a grown man cry!

The meeting took place at a new venue for the Section, and one that proved to be very satisfactory: Nat's Airport Ballroom, 12101 S. Crenshaw, at the southeast corner of the Hawthorne (Northrop Field) Airport. The room is very quiet (only one take-off during the meeting), is spacious, has an open bar, and served an excellent buffet dinner. Hat's off to Program Vice Chair, Bob Conger, for finding it. Hat's off to the Air Force for hangin' in there!

A HYPER EXPERIENCE

The September 18, 2002 Dinner Meeting at the South Bay Grill provided, at $25 per, both a clear look into where we are headed in the development of air breathing Mach 6-12 vehicles and a very good dinner, albeit the dining service for the 68 attendees was too slow.

The featured speaker, Dr. Kevin G. Bowcutt of the Seal Beach - based Boeing gang, accomplished a magical feat of cramming a 3 hour briefing into an exciting 70 minutes. The summary, as I interpreted it, is that the present DOD/NASA long range goals will be very difficult and expensive to achieve, but that the prize is so important that money will continue to be available, even in the face of slow, agonizing progress.

Dr. Bowcutt, educated at the University of Maryland under the tutelage of my friend John Anderson, is a Boeing Senior Technical fellow and Chief Scientist of Hypersonic Design and Applications. He is an expert in the theory of flight of hypersonic vehicles. He is currently leading the design team for third-generation reusable launch vehicle (FASST) at Boeing, and the hypersonic vehicles portion of the Boeing Integrated Vehicle Design System (BIVDS) project. He titled his talk: *"Toward Practical Long-Range Hypersonic Transport and Routine, Affordable Space Access"* – a long but appropriate title. He noted that even though his work is at the very leading edge of technology, Theodore Von Karman had long ago, in March, 1944, opined that the use of the oxygen in the atmosphere as a propellant was clearly the way to go for accelerating to hypersonic speeds. Boeing, DOD, NASA and others in Europe, Russia, and Australia seem to agree.

The immediate aim of Boeing's program is to investigate three application areas: High speed Mach 6-8 cruise missiles; Rapid global transportation – both commercial and military; and routine low cost access to space. His talk concentrated on the first application, and has adopted the philosophy that this can best be accomplished using an air-breathing engine with supersonic combustion features. The aim of the harder third goal is to attain space at one hundredth of present costs. In sight are the winning numbers for the second: at Mach 6, New York to Paris – 55 minutes; L.A. to Tokyo – 1.4 hours!

He pointed out that vehicle designs being considered have the following features:

- They are slender for low drag, operate only at low angles of attack, and have sharp leading edge swept back lifting surfaces
- A large amount of lift is developed by the flow of air through the propulsion system, and indeed this makes for control and balance problems. Also, at high speeds, an increasing fraction of lift comes via centripetal acceleration and resulting centrifugal force.
- The vehicle must be highly integrated – there is no separating the airframe from the propulsion system - the inlet makes up an appreciable portion of the frontal area and the exhaust is takes up most of the vehicle base area. These essentially 'design' the vehicle.
- There are severe heating problems – temperatures above the 10,000 degree-F range have to be dealt with in stagnated air regions
- Presently, until more sophisticated design analysis and optimization tools are developed, design optimization is difficult and there is much uncertainty in designs developed

He next discussed an aerospace plane that would ply the Paris and Tokyo routes. First, the question of fuel: they have looked at LH2, hydrocarbons, and JP-7. Hydrogen looks like the only real choice, but its selection leads to larger vehicles due to its poor volumetric efficiency even in the liquid form. Then there is the decision of horizontal vs. vertical take-off. Finally – a problem that all hypersonic vehicles face – how to get up to hypersonic speeds? Choices being considered range from turbojets to air-augmented rockets. Both separable boost systems and integrated ones are being looked at. Considerations are turnaround time; gross take-off weight; expense (which may favor pure rockets!); the possibilities of performance growth – which favors air-breathing technology. DARPA is sponsoring a system study called "HyperSoar" that is focused on design trade-offs leading to the development of technologies and vehicle concepts needed to meet the goal of rapid global reach.

Boeing, of course, is not alone in these studies. Around the world, there is considerable action:

- Russia worked on a liquid hydrogen scramjet, boosted to speed by a missile
- A second Russian program, designated "Eagle", is a larger ballistic missile-boosted scramjet development
- Rockwell used a light gas gun to boost a 4-inch scramjet test article to Mach 8, with modest success.
- GASL of Long Island followed Rockwell (now Boeing) in developing a hydrocarbon-fueled, light gas gun launched model up to 8100 fps
- There is an Australian scramjet flight test program using hydrogen fuel which reached Mach 7.6 on a ballistic reentry rocket-boosted flight
- DARPA/ONR are sponsoring the "HyFly" program, aimed at flight testing a dual combustion ramjet (DCR, designed by JHU/APL) on a Mach 6 demo missile
- NASA/DOD is developing an X-43 series of flight test vehicles under the Hyper-X program aimed at achieving operational flight vehicles by 2020. Flight demos at Mach 7, 10 and even 15 are planned. The first vehicle, presently under test and already a victim of booster failure, is boosted by Pegasus and is 12 feet long. X-43B has two candidate versions, both about 40 feet long. One will be selected

and launched from a B-52 circa 2010, powered only by its own air-breathing engines (i.e., no booster). X-43C will be boosted by a Pegasus to Mach 5, but will then use an Air Force/P&W JP-7 scramjet to accelerate to Mach 7.

In summary, Kevin thought that what is really needed now are improved analytic tools to allow very sophisticated trade studies and vehicle optimization; and, a National Aerospace Initiative to coordinate and sponsor existing and future Air Force, Navy, Army, DARPA and NASA hypersonic programs.

As a listener and note taker, I came away with the sinking feeling that even though positive thrust has been detected (they think) in some of these new propulsive systems, we have a <u>very</u> long way to go. The problems seem immense, the pitfalls many, and the lack of adequate testing methods obvious. Nevertheless, can you imagine the excitement of working in this field? Makes one wish he weren't retired!

A GEM IN THE WEST

We Angelenos, who should know better, have a tendency to take the Jet Propulsion Laboratory for granted. We shouldn't! The Lab is undeniably the world leader in planetary exploration and the cutting edge technology that goes with that enviable position. This point was vividly struck home by JPL's Deputy Director, USAF Lt. General (ret.) Gene Tattini in his dinner meeting talk to the Section on November 21, 2002.

A brief historical record of JPL achievements was the first area addressed by General Tattini in his presentation entitled, "Space and Earth exploration: Our Vision for the Next Decade". He noted that the Lab was founded in 1936 by von Karman and his Caltech cohorts and was transferred to NASA oversight in 1958. JPL led the development of US rocket technology in WWII. The General, now in his 16th month in his 'new' job, has excellent oversight of the technical activities and the political battles that the Lab is continuously waging. Gene came to JPL after a distinguished career of 33 years in the Air Force, culminated by his directing the AF Space and Mission Command (SMC).

The remainder of his talk covered JPL/NASA goals and objectives in expanding their traditional role of deep space exploration to new partnerships with DOD, Industry, and internal security government agencies. He explained that the former NASA "Faster, Better, Cheaper" credo allowed JPL to break out of the mold of one very big spacecraft every so often into their present multi-faceted mission program, which now is fielding 14 spacecraft operating across the solar system. They also have collaborative efforts with programs of NASA/ Goddard, NASA Ames, and the Applied Physics Lab. The 14 operational JPL spacecraft consist of 2 *Voyagers* on interstellar missions, *Ulysses*, *Genesis* solar wind sample return, *and ACRIMSAT* studying the sun, *Galileo* and *Cassini/ Huygens* studying Jupiter and Saturn, *Stardust* returning comet dust, *Mars Global Surveyor* and *Mars Odyssey* in orbit around Mars, and earth-monitoring *Topex/ Poseidon, Quicksat, Jason 1, and GRACE* earth gravity measuring mission.

In addition to these ongoing programs, JPL has 7 or more missions in development. These include the third of the great telescopes, the *Space Infrared Telescope Facility* (SIRTF), the *GALEX* ultraviolet observatory, 2 *Mars Exploration Rovers* (MER-1 &-2), a deep impact probe, and special instruments mounted on EOS-AURA: The Microwave Limb Sounder (*MLS*) and the Tropospheric Emission Spectrometer (*TES*), and *Cloudsat*. In particular, January 2004 will be a banner month, featuring 2 launches, *Stardust*'s encounter with Comet Wild-2, and 2 Mars rover landings. Stand-by!

The General devoted significant time to the overall international Mars Technology Program. Its goal is to "Establish a permanent robotic presence on Mars to study the geologic and biological evolution of our sister planet". This field is called 'Astrobiology' and its theme is "Follow the Water" utilizing orbiters, landers, scouts, and return vehicles. The latter systems include rendezvous and sample capture/return-to-earth spacecraft along with a Mars ascent vehicle. Surface mobility, subsurface access and in-situ life detection instrumentation devices are required. In 2003 the ESA *Mars Express* and the Japanese *Nozomi* orbiter joins the 2001-launched NASA *Mars Odyssey* which will deliver the Mars Exploration Rovers. In 2005, the NASA *Reconnaissance Orbiter* arrives, and in 2007, two new orbiters – the ASI-NASA *G. Marconi Telecom Orbiter* and the French *PREMIER- 07* science orbiter, arrive. Also, scout missions utilizing French *Netlanders* will commence. In 2009, the ASI-NASA SAR science orbiter will be on station, having delivered to ground the NASA *Smart Mobile Laboratory*. Plans for the following decade include more orbital recon, in-situ life detection, sample return missions, subsurface access, and in-situ exploration.

Hand-in-hand with these programs is an upgrading in the technology and engineering knowledge in key areas critical to deep space exploration, led by Lab personnel. Targets are autonomous mobility; deep space communications advances, featuring optical downlink frequencies to open bandwidth; improvements in deep space navigation with highly stable clocks; extreme precision formation flying for science and rendezvous; high precision instrumentation in optical to sub-millimeter frequencies, including interferometry; and application of active sensors for mapping and positioning using SARs, altimeters and local GPS systems. I noticed that there was no talk of 'Manned Mars'. Obviously, such adventures are twenty or more years away.

The Jet Propulsion Laboratory physical plant, personnel utilization, and budget are prodigious. The Lab has a $1.4 billion business base; hires 5400 people including about 500 on-site contractors, is located on 177 acres with 134 buildings and 57 trailers. The 860,000 square feet of lab spaces exceeds office space by almost 200,000 square feet. The biggest share (38%) of expenditures support planetary flight projects, with 20% devoted to astronomy and physics and 16% to Earth Science and Technology. 13% supports the interplanetary network and 10% goes to solar system exploration. 55% of the staff works in R&D and 15% in R&D management. The 2900 R&D staff is almost evenly divided by thirds in PhDs, Masters or Professional, and Bachelor degrees.

Before I close, let me muse about my personal memories of some of the great things the Lab accomplished prior to the space age. My first brushes with them occurred in the atomic age. Therein, the Corporal missile, developed by JPL and Homer Joe Stewart for the Army, was the first missile to incorporate the atomic warheads that I was working on at Sandia. In the early 50's, we tested special barometric fusing devices for the missile in Bud Schurmeier's multi-speed supersonic wind tunnel. A decade later, JPL space pioneer Jack Froelich, a principle in our country's first satellite, Explorer I, became V.P. of my new company, Space-General, and led us Aerobee rocketeers into the newly burgeoning space-age. We soon successfully flew OV-3, the Air Force's third scientific 'Orbiting Vehicle'. Space–General was an offshoot of Aerojet, which itself was spawned by JPL as the US's first rocket firm.

Around 70 enjoyed the General's briefing at the Proud Bird restaurant. One always holds one breath there, waiting for a landing behemoth to fly into the window, but somewhat surprisingly the noise is pretty well under control. On the other hand, the roast beef, or whatever is was they served as an entree, was tough and sinewy. I've spent 25 bucks for better food.

THE MISSING LINK

A main objective of the Air Force Space Command is to obtain the capability to surveil ground and ocean areas throughout the world at all times, and to integrate all such information sources in a way that interpretation and reaction can be accomplished in a timely manner. The latter job is as difficult as the former. The Space Based Radar program, now in its relative infancy, is a vital lynch-pin part of this grand scheme. Somewhat surprisingly, I helped propose on the first study of a U.S. SBR system just before I retired from TRW in 1988. Recently, however, it has become a bona fide funded program with a high priority. *"Space Based Radar - A System and Program Overview"* was the subject of the January 15, 2003 Dinner Meeting briefing as presented by Colonel James Painter, the SBR Program Director at SMC.

Since October 2002, Col. Painter has had direct responsibility as the USAF lead, for the SBR Joint Program Office (others involved are the National Reconnaissance Office, the USAF's principle partner in the program; the AF Research Lab at Kirtland; DARPA; and the Electronic Systems Command at Hanscomb Field; along with assistance from NASA) to provide the capability required to support the national war fighting strategy. Jim entered the Air Force in 1978 having graduated from the University of Pittsburgh ROTC program. He was promoted to his present rank in September, 1998, after a distinguished always-space-oriented career as a space systems user, developer, manager, and analyst. Along the way, he was the Deputy Director of Requirements at Headquarters, Space Command in Colorado Springs.

As intimated above, SBR – after a long legacy of study and waiting for the technology to develop, is finally a program high in the priority of the Pentagon and Congress, and Jim must closely follow the rise and fall of his budget as the debates continue. One of the primary purposes of the SBR is to detect and identify strategic moving targets on the ground. That technology will soon allow for this was confirmed last February (2002) by the issuing of the results of a seminal summer study from Lincoln Labs. As a result, the Joint Program Office was established and a significant budget assigned. Sources estimate that as much as 300 million have already been spent to date on past incarnations of this capability, including the recently cancelled Discoverer program.

The Colonel pointed out that our present pertinent capabilities are inadequate in coverage, poorly integrated, untimely in reportage, and cry for unconventional approaches, unquestionably utilizing sophisticated computer networks, to achieve the ultimate aim. This aim calls for an integrated system network that can accept inputs from all intelligence gathering and surveillance agencies – military and civilian. What we get when the goal is achieved is precision warfare, a strong deterrent, rapid targeting, and improved homeland defense.

What his program office is now wrestling with is addressing the architectural integration of allocated SBR resources to the proper theaters, getting the data to where it is needed strategically, tactically, and nationally. Before forging ahead, many trade-offs must be made to assure that the SBR constellation can perform its three main tasks: Imaging, terrain profiling, and detecting ground moving targets. The preliminary thinking calls for a mix of 9 spacecraft deployed in its first developmental spiral.

The conceptual program schedule calls for completion of requirements and risk reduction studies by '04; The risk reduction studies will concentrate on the space and ground architectures as well as integration with current and proposed Theater BMC3I and National infrastructure. Already under contract are studies on the ESA (Electronic Scanning Array), now concentrating on a 16x3 meter phased array; on system concept development; information management; integrated payload design; on-board processing; and spacecraft technology. RFPs are being issued for other study areas. Col. Painter invited the audience to share ideas and ensure the maximum contact between the government and industry regarding the program.

The meeting, at the South Bay Grill, was attended by 48 people and my generous portion of Cod was delicious, though clearly outside of the boundary of my recent promise to my primary care physician that I would lose 10 pounds before I next saw him. The other choice was a chicken dish which looked equally appetizing. Vice Chair / Programs Bob Conger of Microcosm arranged for and introduced the speaker. Good Meeting!

TOUGH TIMES IN THE GLOBAL LAUNCH INDUSTRY

Market projections made by analysts and industry groups in 1998 indicated that there would be 55 medium-to-heavy launches of commercial satellites in 2003, plus a comparatively small, stable government demand for launch services. Back in 1998, the preceding five years had experienced a 270% increase in commercial launch rates, which lent credibility to these market projections. No wonder investors entered the fray with such a vengeance, believing they could position their products to capture a healthy share of a promising growth market.

Today's reality is that current predictions are for only 18 medium-to-heavy commercial satellite launches in 2003, while the available launch capacity for such launches has grown to about 120 worldwide! And, to make matters worse, a launch market recovery is not expected until '07-'10. Venture capitalists who helped fund both low Earth orbit constellations and rocket development start-ups back during the boom, unfortunately, are left with little but write-offs and assets sold at pennies on the dollar. However, key launch providers, including Boeing, will be able to weather the storm, thanks to continuing government (NASA & DOD) business. In the wake of this commercial downturn launch providers are struggling to remain profitable even as new systems continue to be introduced around the world.

Jayne Schnaars, Vice President of Boeing Launch Services of Huntington Beach, the speaker at the February 20, 2003 Dinner Meeting, compared the 1998 projections with today's situation, showing current and near-term demand is only a fraction of previous projections. About 45 attended her talk at the South Bay Grill, entitled *"The State Of The Launch Industry"*. Jayne focused on a discussion of the launch market, which today includes both commercial (communications and remote sensing) missions and government launches.

Ms. Schnaars, with Boeing since 1986 in various program management and business development capacities, was appointed to her new assignment last October. She was most recently the Division Director of business development for Boeing's human space flight business unit. She has a BS in applied math and statistics from the State University of New York and an MS in mathematical statistics from Ohio State University. Her present responsibilities include sales and marketing of the Sea Launch and Delta family of launch vehicles.

The Delta family is nearing its 300th launch. NASA's workhorse for science missions, the Delta II, has a remarkable 98% success rate (103 out of 105), and was recently awarded 19 additional launches through 2009. The Delta III is a higher performance variant of the Delta II. Boeing recently introduced the much larger Delta IV, built for the USAF's Evolved Expendable Launch Vehicle (EELV) program. Delta IV has 5 configurations, including the Heavy, which has a >50,000 pound lift capability (about the same as the Space Shuttle) to low Earth orbit. The total range of lift capabilities to GTO across the entire Delta family of launch vehicles is from approximately 2000 to 29,000 pounds.

The successful Delta IV inaugural launch took place on November 20, 2002. The vehicle features a newly developed engine, the RS-68, and has been awarded 22 of the 29 EELV launches currently contracted by the U.S. Air Force. It can be launched from both the Cape and Vandenberg. In response to the government's allocation of $500 million each to Boeing and Lockheed Martin in partial support of the EELV program, Boeing built a new plant in Decatur, Alabama with a capability of producing up to 40 common booster cores per year. A special 'Delta Mariner' ship can carry the vehicles from the Boeing Decatur factory to the Cape in 14 days and to the West Coast in 28 days.

To date, Sea Launch has completed 7 successful launches, and has an additional 17 commercial launches awarded. Five of these are scheduled for 2003. The Sea Launch system is a 3-stage Russian/Ukrainian 'Zenit' launch vehicle with a GTO capability of 6 metric tonnes. The launches are made from a site in the Pacific on the equator, near Christmas Island.

Looking to the future, Boeing has a robust launch manifest of 18 launches in 2003 and 21 launches in 2004. In the bigger picture over the next five years, they anticipate 10 Sea Launch, 22 Delta IV, 7 Delta III, and 42 Delta II launches. So, barring serious set backs, Boeing Launch Services' near term launch base seems assured.

There is an abundance of transponder capacity already in-orbit. In 2001 there were 24 new commercial satellite orders; in 2002 only 3! The current worldwide forecast of launches through 2010 shows an average of 5/year to LEO/MEO, and 20/year to GEO, with a partial recovery expected in the 2006-2008 period. And this while the world's GTO launch capability is 70 launches per year (made up of 6-7 by Sea Launch; 17 Delta IVs from the Cape and 9 from Vandenberg; 10-15 Protons from Baikonur; 8-10 Ariane 5s from Kourou and 12 Atlas 5 launches from the Cape, not counting additional capability provided by Japan's HII-A, China's Long March and India's GSLV! Another problem in the commercial sphere is that the space insurance business is having difficulty maintaining long-term profitability and rates are increasing.

In the U.S., the Air Force will continue to support two major launch vehicles through its EELV contract. Boeing is currently evaluating the possibility of a Delta IV Heavy launching an Orbital Space Plane to the International Space Station.

Who's the villain in this fiasco of lowered expectations in the commercial Comsat build and launch biz? The success of fiber optics networks is one possibility. But, who knows – it's like trying to predict Wall St! Or maybe it's me? As I left for the meeting, my wife badgered me on how come I could spend 25 big ones for a dinner, when our usual weekly 'date' night out consisted of dining for two at the Sizzler or McDonalds? When I got home I told her that she wouldn't have enjoyed the sumptuous trout entrée anyway because the portions of everything were so huge.

WHO'DA THUNK IT ABOUT "SKUNK"?

Sometimes volunteerism gets a soul into trouble! Who'da thunk that I, a man of many words, would have trouble writing up a review of the Section's 2nd Dinner Meeting of the

Fall season. But, my friends, telling you about the talk presented by Jeffrey "Skunk" Baxter on October 22, 2003 at the LAX Crowne Plaza, entitled *"How Did a Rock Star Become an Advisor to the Intelligence Community?"* was indeed a challenge. How do you say? - an enigma wrapped in a message or messages which I shall try to reveal.

The first part is easy. I asked the speaker, a lead guitarist who was a founding member of the group 'Steely Dan', and later with the equally celebrated 'Doobie Brothers', how he got his sobriquet. Right away I could see potential problems when he said, "Ah, you'll have to read about that in my soon-to-be-published book". But after that, it becomes more difficult as he explains away his present dual life in music as a performer, record producer, and a composer and his other activities as an advisor to the government. He is in fact a consultant to the Missile Defense Agency, to the National Mapping and Imagery Agency, to Congressman Dana Rohrabacher, to several other government agencies, including the NRO, and to aerospace companies interested in technologies for future warfare. Additionally, he works with local police agencies on matters of early warning and anti-terrorism.

How did he get that way? He had a great interest in using high tech advances to apply to guitar and recording technologies and, early on, voraciously read '*Aviation Week*" and "*Janes*" to look for application crossovers. In 1991, he wrote a paper, "*How to convert the Aegis System to Theater Area Defense*" which caught on in Congressional and military circles, and the ride began. He was appointed to the Civilian Advisory Board for Missile Defense and gradually found himself in demand as he developed a 'kick ass and raise hell' style of thinking differently that they liked. He said that he's now been in the program for 11 years asking, "Is there a solution for this problem" and playing the Devil's Advocate on 'Red Teams' to get his ideas across. Although he is a poseur at being hip technology-wise, it is obvious that he has done his homework and knows what he is talking about. And, he did a very funny bit on having us imagine his and the FBI's problems obtaining his top security clearance. As for his musical skills, I leave that to the great unwashed who do not share my moldy fig musical tastes.

But, the important part of the evening was his message. He is a very strong advocate of expanding the breadth of back grounds of the people responsible for our defense future and of using very innovative approaches by 'thinking differently'. He argues that using celebrities who have developed other skills and ideas, perhaps even outre ones, to attack some of our problems is the way to go. "Expand the labor base!" He cited that some unusual skills developed by Dan Ackroyd and James Woods were already being taken advantage of in addition to his own contributions.

He recited some of his present concerns; The fact that the Chinese are moving forward at a rapid pace and we are sitting on our hands; the fact that we are losing our space impetus and not exciting the minds of the young sufficiently; and, worst of all, we are not being smart about what we should be doing. He summarized his thoughts by making the following pleas:

- We must excite young people to get involved in technology. The AIAA should be a player in this.
- We must 'think differently' to achieve new goals.
- We must find diverse, unusual people to make this happen.
- We must create new connections between these diverse sets of people
- We must find people willing to take chances and give them a chance to fail.

Rhetorically, he asked, "What is the solution?" He said, "NASA is not the solution, Private Business is!" As an illustration he offered the example of the Russian willingness to take private parties into space for a fee. More adventurism like that is what he advocated. He again advocated reaching out to artistic people, saying that, "Washington is Hollywood for ugly people". He dissed both Presidents Clinton and Jed Bartlett, making me believe he was of a conservative political bent, which to my mind is NOT thinking differently. It was a very interesting and amusing evening - and the food was good, but the helpings too big.

ANYONE FOR ROCKET TESTING?

Looking for a nice, environmentally friendly, well equipped place to test your favorite rocket engine, motor, or associated component? Well, my friends, Colonel Joseph F. Boyle, Commander, Air Force Research Laboratory at Edwards Research Site, is just the man to see - bring money and/or an approved government testing need. I know this because I was one of approximately 70 listeners who drank in his words at the January 21, 2004, Dinner Meeting at the airport Crowne Plaza hotel. His talk was formally entitled, *The Future of Space Launch and Propulsion's Role* and, in it, he described his facility and its current support and research efforts.

Col. Boyle is enjoying his fourth assignment at Edwards AFB, where he is Associate Director of the AFRL Propulsion Directorate. In his previous billets here and at SMC and in Germany, starting in 1979, he has worked as a project/program manager in programs dealing with rocket propulsion, launch systems, manufacturing technology, and structural analysis. He is a mechanical engineer by training and has an MS in Science System Management from USC. In his present job, he is not only responsible for obtaining and overseeing test work for his Lab, but also has taken on the added task of improving coordination between the US's military and civilian rocket testing organizations.

His present bailiwick, in the Eastern corner of the huge Edwards Air Force Base, is a 65 square mile compound – which he noted would encompass Washington, D.C. – with 135 major labs, 30 major test areas, and approximately $2.5 billion in facilities. He is responsible for the safety and security of over 575 people working at 'The Rocket Site'.

His facilities have six basic rocket test capabilities, plus a highly capable chem lab:

- LOx/LH2 up to 1.5 Mlbf thrust
- LOx/LH2/HC (hydrocarbon) to 1.5 Mlbf
- Vertical SRM (Solid) up to 3 Mlbf
  LOx/HC to 1.5 Mlbf
- SRM horizontal (as needed)
- Satellite propulsion at 100Kft altitude
- Upper stage propulsion at 100 Kft altitude

The colonel pointed out that the latter two facilities did not compete with similar facilities at AF Arnold AFB where routine check-out high altitude proof firings are made. The facilities at Edwards are employed for high risk and first time testing because of the remoteness of their location.

Their current work load is mostly in direct support of Air Force Space Command efforts on 1) Land based strategic deterrents (e.g. new ballistic missiles); 2) Operationally responsive space programs, especially FALCON (Force Application Launch from CONUS) – a program of developing responsive LV's and TacSats (now looking at small, 1000# payload quickly launched LVs and responsive spacelift vehicles of the EELV family); 3) DARPA ('Rascal' program) and DDR&E projects; and 4) Support of NASA's Shuttle return-to-flight program.

A major on-going effort is called IHPRPT, which somehow they are able to make a word out of (like Joe Btfsplk of L'il Abner fame). In this case, it means 'Integrated High Payoff Rocket Propulsion Technology'. Its aim is make improvements in all rocket figures-of-merit and is a coordinated Air Force program with the Army, Navy, NASA, and DDR&E. It is supported by ARC (now Aerojet), NGST (was TRW), Aerojet, Boeing, Alliant (was Thiokol), and Pratt & Whitney. Its goals – to be obtained by 2010 - are concentrated on improvements of boost and orbit transfer propulsion, spacecraft propulsion (including electric, mono-props, bi-props, and solar/thermal) and tactical propulsion.

Their major on-going research programs are concerned with high energy density matter (Polynitrogen – a possible 'heavy' replacement for LH2); Ionic liquids (zero vapor pressure replacement for hydrazine); new hydrocarbon-based fuels (higher energy); and a variety of electric and plasma thrusters.

The test site has a long legacy of support to the Air Forces key programs; Minuteman, Peacekeeper, and the Titans, Atlas's, Thors and Deltas of yore and present. The facilities are available to all performing government work and to private industry by arrangement. "Try us, you'll like us", sayeth the Colonel.

And, to make me eat crow about only getting served chicken, the fare was roast beef this time, and quite good!

(HEIGH-HO)^2 - IT'S OFF TO MARS WE GO !

The final Dinner Meeting of the 2003-4 season took place on May 20, 2004 at the LAX Crowne Plaza, and featured JPL's Dr. Mark Adler, who is playing a key role in the mission of the two on-going Mars Exploration Rover vehicles, 'Spirit' and 'Opportunity'. The 42 attendees were treated to an inside look at both the preparations for and the operations of the two essentially identical MERs. The bottom line is a source of great pride for all involved: The mission purpose – that of determining whether liquid water ever existed on Mars – has been decisively proven in the positive. "Lots of it, and deep"! This having been proven by the Opportunity, the next missions will be aimed at finding positive evidence of the prior existence of any identifiable life forms.

Prior to his talk, several bits of present and future business were accomplished. Jim Wertz, - in his swan song as L.A. Section Chairman, was presented with the plaque, noting his ascendancy to the coveted FELLOW grade, which he would have received at the National Awards Banquet in Washington in April. This event took place at the same time as the local 2nd Responsive Space Conference, which he chaired. Jim announced that the 3rd responsive Space Conference would take place April 19-21, 2005. Write that down!

There were a number of student awards presented by Section Vice-Chair Jane Hansen and the Professors who guide the AIAA student branches at USC, UCLA and Cal State Long Beach. The Universities each named two outstanding students, all of whom received 500 dollar awards. Three undergraduate awards of plaques and money were granted to Honeywell Science Fair winners whose projects were aerospace related. A special $500 award was also presented to the UCLA coed who organized the recently completed AIAA Regional Student Paper Conference. Hansen praised the student's efforts and noted that we are the only Section with three student branches in our purview.

Dr. Adler, a U. Florida/Caltech graduate, is responsible for the mission system operation of the MER project which designs the mission, develops mission operations systems, trains the operations personnel, and operates the missions while in transit to Mars and on its surface. In addition, he is now in charge of the "Spirit's" operation. He is an experienced 'traveler' to Mars, having been previously involved in the Mars Sample Return Mission. His talk was entitled, "Exploring Mars – *The Journey of the Mars Exploration Rovers*: How the rovers Spirit and Opportunity were designed, built, tested, flown, landed, and driven, how they work and what they've done". In other words, the whole shebang!

The first Mars landing explorer mission, the 'Pathfinder', took place in 1996. Future missions are planned for 2005, 7, & 9. Mark expected that future Rovers would be different for each mission, in particular their instrument suite would change and hoped-for improvements in the landing system would be developed. The present 6-wheeled Rovers stand about 3 feet high and have a main frame of around 3x4 feet, with unfolding solar panels and an upward extending submarine-like 'neck' which hold two sets of aim-able stereo cameras – to provide some depth perception. The 'Pancams' (panoramic viewing) have a 17 degree field of view and the 'Navcams' (navigation) have a 45 degree FOV. The 'head' also contains an IR Spectrometer. In addition, attached to the main frame are front and rear (for back-up) Hazcams (hazard) to aid in operation. A forward robotic arm, with much the dimensions and operations of a human arm, contains an imager, spectrometers, and a rock abrasive drilling and scaling tools. Communications directly to earth are accomplished through a high gain antenna, with an LGA as a back up; while a UHF stub antenna communicates to a Mars orbiter.

Mark then described some of the difficulties in the design and test of the landing package. After atmospheric drag reduces the reentry velocity from about 12,000 mph to 1200 mph, a parachute system, further decelerates the landing package to about 200 mph. Pioneer supplies the decelerator, whose design was arrived at after much wind tunnel and drop testing. Finally, rocket deceleration is employed close to the surface, but the last 10 meters are a free fall drop. The clustered air bag device surrounding the payload protects it from rocks as it bounces merrily on the surface. The air bag system was tested in a very large vacuum facility, replete with all kind of anticipated rocks. It required many re-designs until a blow-out-immune combination was found.

He said that his baby, the Spirit, had not found evidence of water and was now heading for the hills, where hope springs eternal. On the other side of the world, however, Opportunity has already hit pay dirt and is looking for further corroboration. It found both chemical and geological evidence of the existence of liquid water. It is now circling the lip of a crater while they decide if it is safe to venture downhill where they hope some more good stuff will be uncovered.

As usual, the dinner was good and the portions still too big. They also feature very sweet desserts which I shun for fear of going into diabetes shock.

ARE WE OUT-SOURCING ENGINEERING AND SCIENCE?

This was the major question raised by UCLA Aerospace and Mechanical Engineering Department Professor, Dr. Ann R. Karagozian at the March 23, 2005 Dinner meeting at the LAX Crowne Plaza hotel. Her talk, entitled, "*Perspectives on Aerospace Engineering in the USA*" asked several intriguing questions, mostly leveled at 'aeronautics' – as opposed to 'astronautics' - which she attempted to answer:

Are we moving in the right direction –especially budget-wise?
Are we laying the proper foundations in education?
Are we protecting our future?

It was the latter question that most concerned her, in view of a downtrend in research funding; a lack of strategic vision, and the possibility of events overtaking preparation. Dr Karagozian, an AIAA Fellow, who is Head of the UCLA Combustion Research Laboratory, cited our continuing need for bright students to enter the aerospace field, faculty who can train them in the latest technology, and young working engineers gaining on- the- job experience.

Her fear is that because Americans are viewing engineering and science as perhaps the poorest paid field of the higher echelon employment opportunities, that some or much of our high tech R&D will be out-sourced to the countries whose young engineers and scientists are – in ever increasing numbers – being trained here at US universities. It is apparent that more monetary rewards can be had in lawyering, doctoring, MBA-ing, and –in California, at least - in Real Estating!

She notes a decline in US citizens in training for R&D; a drop in our production, vis-à-vis other countries, China and the EU in particular; a fear of our losing dominance in engineering and science. A great deal of our research funding is going to the NIH – to support medical research. This, in spite of there being job openings and an increasing R&D work force need in engineering and science here in the US.

Dr. Karagozian suggested a few ways to improve the outlook:

Provide more incentives to pursue the PhD and careers in R&D

Re-think the NASA budget to siphon off less money for ISS, Space Shuttle, and other big ticket money eaters

Put more money into aerospace R&D

GPS THEN, NOW, AND THE FUTURE

Although Col. Rick Reaser has been heavily involved in the GPS program for a total of 12 years, his wife never appreciated what he was doing until their car was recently equipped with city map and talking lady equipment. "She fell in love all over again", so said Rick in his opening remarks at the January 18, 2006 Dinner Meeting to a crowd of about 60.

Col. Reaser, an Air Force Academy and Naval Postgraduate School graduate now on his third Global Positioning System Satellite stint, is Deputy System Program Director of the Navstar GPS Joint (i.e., International) Program Office at the local Air Force Base He was previously Chief Engineer in an earlier assignment here. His general topic is shown above as the title of this review, and he presented it in six parts: What is GPS?; What is GPS used for?; Its History; Description of System; How it Works; and Its Future. The audience greatly appreciated his complete in-depth knowledge of the system, and badgered him with questions long into the night. I learned many interesting facts, and will cover some of them in what follows:

There are two levels of GPS usage: Military and Civilian, and three segments comprising the system: Ground (i.e. control), Space (with around 28 operational satellites as of now), and User. The GPS system tells the user many things, such as latitude/longitude, speed, average speed over a course, elevation, how far you've gone and how long it has taken, estimated time of arrival, path traveled, plus many more innovative usages. The 'smarts' of the system are all in the individual receiver, and are better and more accurate the more you pay, starting at slightly over $100. He listed three Military usages: Navigation, Precision Control of Weapons, precision timing; while there are a plethora –he listed about 15 - of well known civilian utilizations. The aforementioned auto navigation is an obvious one, as is usage by sailors and pilots. But the idea of "games", using GPS was new to me. Such games are variously called "Pac Manhattan" (finding people), "Geo-Caching (hiding things), 'Way Marking' (physical location), and there are more being developed every day. A short movie showed another very important usage: The ability of GPS to mark out where a road is, for use by a snow plow in a white-out, for example.

The Colonel then discussed the unusual history of navigational tools culminated by the GPS. The importance of the "Ship's Clock" was brought out, since accuracy of knowing your location is a direct function of how well you know the correct time. LORAN was a precursor to GPS, whose first manifestation was the TRANSIT satellite of the early '60's (Ed. Note: The writer worked on it while at Aerojet/Azusa; who developed the AbleStar upper stage LV – the first re-startable upper stage which placed TRANSIT in its correct orbit). It was used by fleet submarines to know where they were. The GPS itself was born in 1973. In 2005, the first modernized Block IIR was orbited in a truly international program with over 40 countries participating.

There have been >50 GPS satellites launched to date. The GPS constellation consists of six planes with four satellites in each plane. Each plane is inclined 55 degrees and orbits are circular with an 11,000 nmi. semi-major axis. They are hardened to operate in the Van Allen belts. Actually there are spares in each plane, allowing the control system to power down satellites for maintenance without loss of system coverage. The average life span has been ~ 7.5 years. GPS satellites have been launched on the Delta II LV and are transitioning to the EELV.

The GPS system is controlled from Schriever AFB in Colorado, with a back-up station in Gaithersberg, Maryland. Support facilities are spread out over the world; Diego Garcia Island, Hawaii, Kwajalein being some of the present support links, with many stations more being added at higher and lower latitudes. In the user segment for the Military, there are a great variety of receivers for various applications, as well as two handheld portable types for the troops; one such being a combat survivor unit. There is also a small module suitable to connect to a lap top computer. In addition, there are 7 or 8 civilian application receivers.

The system depends on each satellite knowing precisely where it is at a time derived by on-board atomic clock synchronization. They report their position and the receiver does the calculation. It is necessary to get signals from at least 4 satellites to get information to locate yourself to about 10 feet. If more satellite signals are received, an improvement in accuracy is achieved. Satellite time/frequency units are accurate to about one second per billion years. Satellites are synchronized to about 10 nsec. accuracy. Users do not depend on an internal time/frequency standard since these tend to be large, complex and expensive. DARPA is working on a "chip-scale" atomic clock that may change this. In addition to the GPS function, they use their excellent visibility of the entire Earth to provide nuclear treaty monitoring.

GPS performance is degraded by spurious delays in the passing through ionosphere and troposphere, signal multipaths (especially in downtowns of cities with high buildings), and the number of satellites visible. The Military systems partially defeat these problems by using more than one carrier frequency. The current GPS receivers being fielded for military use include an "anti-spoofing module" and perhaps a jammer-nulling capability.

The future for the GPS system is bright, although there is reluctance at times in Washington to provide funds for improvements because the present system is so good already. In the immediate future, there will be new military bands added with special anti jam features, as well as a second civilian band These will appear in the Block IIF satellites to begin launching in 2009. Along with them will come a new control segment as well as a new user segment. Further down the road are the Block IIIA which may have new orbital planes, even higher security and anti-jam features.

I'm betting that Col. (or perhaps General) Reaser will be back for his fourth tour to see this day!

THE MISSILE DEFENSE AGENCY HAS A PLAN!

And a mighty good one it is shaping up to be, according to Col. Christopher E. Pelc, the featured speaker at May 17, 2007 Dinner Meeting of the LA Section of AIAA at the LAX Crowne Plaza Hotel. The meeting was attended by over 90 members who were presented with a broad scope look of the 10 year or so planned evolution and capabilities of the DOD integrated multi-layered ballistic missile defense system, initiated in 2004. The key word in the preceding sentence is 'integrated' – for the creation of the Missile Defense Agency at the highest level is permitting allying and working in joint concert all the pertinent elements of the military agencies engaged in Defense - be they land, air, sea, or space assets. And, along with this comes the reality of them being able to talk and react to one another in the same language and in real time!

This reassuring message was delivered by an Air Force Officer who plays a key role in this grand scheme of things. Colonel Chris Pelc is the Program Director for the Space Tracking and Surveillance System (STSS) - Program Office, SMC, Air Force Space Command, LAAFB. STSS is the updated reincarnation of what you sports fans used to know as 'SBIRS-LOW' and is a major keystone player in the evolving BMD firmament. In 2001, SBIRS-LOW was transferred to the Missile Defense Agency (MDA), and in 2002 was renamed as STSS. The STSS program is a space-based sensor component of the Ballistic Missile Defense System. The program uses sensors capable of detecting targets utilizing visible and infrared bands. It will consist of low-orbiting satellites designed to detect and track ballistic missiles in all stages of flight. Data from STSS will allow interceptors to engage enemy missiles as early as possible in their trajectories and discriminate between warheads and decoys.

Col. Pelc is the Program Director for the Space Tracking and Surveillance System (STSS) Program Office. He is responsible for the development and program execution of the STSS research and development program. The $4 billion STSS program is an element of the overall Missile Defense Agency's Ballistic Missile Defense System. Col. Pelc graduated from the University of Notre Dame in 1981 with a Bachelor of Science in Chemistry and began his career as an Acquisition Officer in the F-16 System Program Office, Wright-Patterson AFB, Ohio. While in Dayton, he also received a Masters of Business Administration from the University of Dayton. In 1987, Col Pelc transferred to the Global Positioning System Joint Program Office, Los Angeles AFB, California. After several key assignments and upon completion of his studies at the Air War College in 2000, he became the Deputy Director of the STSS Program. He has served in that role until 2004 when he assumed his current position.

The Colonel stated that MDA's plans for defense against ballistic missiles had a three pronged scope:
Provide an integrated layered defensive structure
Defend the United States, Allies and deployed forces
Develop global access and response
By "integrated" he meant that all missile defense assets – in space, on the land or at sea – will be operated together with all our armed services participating. He identified North Korea and Iran as the present "baddies". Today, Navy cruisers in the Sea of Japan and the Israeli Pine Tree-Arrow system form our first line of missile defense.

Col. Pelc's primary responsibility is to develop space assets for MDA's Ballistic Missile Defense System to include STSS, the Near Field Infrared Experiment (NFIRE) and a variety of additional space-related efforts (microsats, space-based interceptors, etc.) The NFIRE satellite, used to gather near field high resolution phenomenology data, will assist in development of boost phase intercept systems and also assesses the viability of a laser communications system for missile defense applications. It was successfully launched from Wallops Island in April 2007. While studies for the operational STSS are underway, Col. Pelc said that there are funding gaps that must be faced.

The attraction of STSS for the Missile Defense Agency is that it alone has the potential to provide birth-to-death midcourse tracking of adversary ballistic missile threats launched anytime from anywhere on the globe. It will detect and track sea launches and launches from nations of every sizes and shape. Its infrared & visible sensors have 24/7 all-weather capability. It has no host nation basing issues. To support these objectives, the STSS program office plans to launch, in April 2008, two R&D satellites built by Northrop Grumman Space Technology from hardware originally procured for SBIRS LOW into low earth orbits (beneath the Van Allen belts). Upgraded software for these spacecraft is in development for use in the 2008-2010 time frame. Col. Pelc stated "We must walk before we run."

During the subsequent Q&A session Col. Pelc reminded the audience that he believed that today's ballistic missile defense system will work against some threats. We have had 20 successes in 27 tests. We have deployed a family of radars that can observe North Korean launches and cue interceptors in Alaska and California. This system has been on alert when not used for additional testing. The Navy's SM-3 Aegis system has been outstandingly successful in tests against theater ballistic missiles. The first SM-3 Aegis cruisers have been deployed to the Sea of Japan.

In conclusion, the Colonel stated that while there have been delays, the MDA has made good progress, and STSS is moving toward flight testing in the near future.

Prior to the Colonel's talk, outgoing Section Chair Jane Hansen made a Special Service Citation to retired (but recalled-to-duty) Northrop Grumman Space Technology engineer Tom Nosek. The Citation was for his continued outstanding work in support of the AIAA Region VI Student Paper Conferences. Tom was an Audio Judge in 2004, the Judging Chair in 2005 and 2006, and Technical Judge in 2007.

SBIRS DINNER MEETING BRIEFING REPORT

The February 19, 2009 Dinner meeting at the LAX Crowne Plaza Hotel featured Col. Scott C. Larrimore of SMC, Commander of the Space Based Infrared Systems (SBIRS) Space Group reporting on the status of the program. His Group is responsible for development, integration, and launch of the SBIRS Geostationary spacecraft, as well as sustaining the present Defense Support Program (DSP) space system. These space infrared surveillance systems provide overall missile early warning, tracking, launch location and probable impact determination. He has been in this key position since May, 2008

The present SBIRS system development evolved from a 1985 start seeking a follow on and eventual replacement of the DSP (Defense Support Program) early warning satellite array which started in 1970 and whose last launch occurred in November, 2007. The program formally started in 1996 as an outgrowth of two early precursor investigations whose purpose it was to define a long range solution to the problem of world wide (especially North Polar, where

DSP is marginal) detection of multiple location strategic and tactical missile launches; determining 'friend or foe'; trajectory tracking, and impact location prediction, and providing timely hand-off of such data to defense forces who are capable of nullifying them. Additionally, the space elements will also collect data that will improve overall system performance. The initial SBIRS system, now partially developed, consists of 2 HEO (Highly Elliptical Orbit – apogee over South Pole) spacecraft and 4 GEO spacecraft, plus Mission Control and Ground Stations. Lockheed Martin is the industrial contractor. Control and main ground stations are located in Colorado at Buckley and Schriever Air Force Bases.

It is the Air Force's intention to maintain and nurture (plans to conserve on-board fuel are under consideration) the present DSP array to the extent of its continued ability. However, because the complete SBIRS system is still a few years away, the desire for continued and robust strategic missile warning capability has led to the initiation of one "augmentation" spacecraft which will relieve some pressure on the program.

At SBIRS program start in 1996, there were two space elements: SBIRS-High, then thought of as a DSP follow-on; and SBIRS-Low, a "Brilliant Pebbles" follow-on focused on post burnout tracking. The latter program was taken over by the Missile Defense Agency and expects first launch of a demo system this year. The SBIRS-High program has evolved into Col. Larrimore's task area. The Colonel described some of its features:

• The goal of all the space elements is to be much more sensitive than the DSP sensors and to facilitate pertinent data hand-off.
• Space elements have both staring (at multiple ground targets) and scanning capabilities; the former capability

gained by 2-axis gimbaled highly agile sensor telescopes and pointing mirrors.
- The IR bands utilized are SWIR and MWIR, with STG (space-to-ground) ability.
- HEO 1 and 2 are already in orbit, and HEO 3 and 4 contract negotiations are under way.
- GEO 1 will be launched in about a year, with GEO 2 about a year after that. GEOs 3 & 4 will be cloned versions.
- In addition, studies are being initiated on behalf of future generation infrared system improvements. More sensitive or larger detectors, on-board processing and faster communication through-put are some of the goals.

Col. Larrimore leads a government program team of 107 personnel and over 1400 contractor personnel. His $20 billion space systems portfolio includes the Defense Support Program and the next generation Space Based Infrared System, SBIRS-High. He holds a BME (Aero & Mechanical) from Princeton and an MS in Astro from Purdue. He has studied at George Washington U. in International Affairs, at the International Space University, and arrived at SMC after an Air Force Fellowship spent at IDA (Institute for Space Analysis) in Alexandria, VA.

As I compiled this Chapter and re-edited its contents, a clear message came through: The people responsible for major progress and discovery in aerospace are the Military services and the Jet Propulsion Lab. I will close by citing the mission of the latter, as presented by Gen. Tattini:

JPL ROLE IN NASA MISSION

- To understand and protect our home planet
  Improve scientific underpinning for environmental and natural hazard forecasts in the areas of solid

earth, physical oceanography and atmospheric chemistry (examples: TOPEX/Jason 1, GRACE, EOS payload, SRTM, GNHNSI*).
Contribute to national security through shared technology development and demonstration in areas of mutual interest (Examples: SRTM, GNHNSI).

- **To explore the universe and search for life**
  Search for life in our solar system (Mars Program, Deep Impact, Europa, PTI**).
  Understand pre-biotic environments (Cassini, Titan, PTI).
  Understand the origin and evolution of our solar system and the universe (SIRTF, Mars Program, Cassini, PTI).
  Understand the fundamental laws governing out universe (Gatex, LISA, low temperature physics).

- **To inspire the next generation of explorers as only NASA can**
  Public outreach
  Educational outreach and participation
  Support of student's and teacher's education

---

*GNHNSI = Geophysics, Natural Hazard and National Security initiative (Code Y)
** PTI = Planetary Technology Initiative (Code S)

Chapter 7

FINAL WORKING YEARS

- VISITING PROFESSORING AT THE TECHNION
- THE START OF THE ISRAELI NAVY
- ABET ADVENTURES
- POWER POLITICS AND FUSION
- RETIRING WITH RELUCTANCE AND A LAW SUIT

In 1996, while teaching my "Space Systems Design" course at USC, I began missing words. Not hard, technical Words – but Words that were just a cut above the difficulty of 'The". I sometimes had to ask the class, "What is the word I'm looking for"? I figured I'd better quit while I was still ahead. Strangely enough, after I packed it in, I no longer missed words? Thus ended a working career of 50 years – but with the 'Golden Years' looming ahead. These are some stories from that pre-retirement period in the late 780s and 90s.

VISITING PROFESSORING AT THE TECHNION

In the 70s, I gave a paper at an International Astronautics Federation (IAF) meeting in Europe describing the Iowa State full scale liquid rocket firing demonstration that I had established as part of a laboratory course in my Aero. E. Department. A professor who taught rocketry at the TECHNION (Israel Institute of Technology) in Haifa apparently heard my talk. Shortly after I retired from teaching at USC, I began to explore Visiting Professor opportunities – armed with the two courses that I had pioneered in teaching: "*Spacecraft*

Systems Design" and *"Principles and Techniques of Remote Sensing* (from Space)". We had always wanted to see Israel, and I knew I had a distant cousin who lived in Haifa. I applied for a fellowship there, and–championed by the professor who had heard my talk several years earlier – was accepted for a semester to teach the spacecraft course. We got there for the Fall semester in 1989. My cousin and her husband, Dick, a Professor of Industrial Engineering at the TECHNION welcomed us and brought us into their family and circle of friends. It was a rewarding experience, which we were to repeat, teaching my other course, a few years later.

The TECHNION is often called "The Massachusetts Institute of Technology of Israel". This is clearly true in a prestige sense. All the engineers of any importance wear the old school tie. It is not true in a pedagogical sense. The teaching there, at that time, was strictly theoretical – no one gets their hands dirty. Practical engineering courses, such as the ones I taught, were looked down upon as greasy grind stuff. In fact, knowing nothing about it, they degraded my remote sensing course from the normal 3 credits to 2! As a result, what started out as a class of 30 ended up with only seven who really wanted to learn the subject, when the word got out about class credit. However, I have recently learned that this high-browed approach has slackened – finally backing off from the left-over Germanic tradition - and the course content is now more in line with American thinking. See for yourself at http://support.ats.org/site/PageServer?pagename=home_template

TECHNION graduates are expected to learn the business by working in the trade plus the experience they get in serving in the Israeli Air Force. Such service is mandatory and they usually get a deferment to complete their bachelor's degree work. Many go back to school for a Master's after their service stint. In the case of my Department there, Aerospace Engineering, the industry jobs were with Rafael – a hi-tech mostly military research and development lab – or IAI (Israel Aircraft Industry).

The school is located in a sylvan setting on a high plateau of about 2000 feet elevation. From parts of the campus, the Mediterranean can be seen both towards the East and to the North, beyond the Harbor to Lebanon. The Aero E. Department is located in a high rise with an overhead enclosed walkway connecting it to another building which also house the Department. Originally, the courses were taught in German, but now most are in Hebrew. Students are allowed to take up to two courses–per-semester in English - these usually being taught by visiting Professors from all countries. The basic salary of the Professors, even the Full Profs, is meager. In fact, on my second time around as a Visitor, the start of the semester was delayed because the faculty went on strike in order to bring their basic salary up to the level of the trash collectors! However, most earn considerably more by taking on projects coming from industry. In addition, they are allowed to 'borrow' from their retirement funds. In the end, this make most of the professors at the high end of the income scale, albeit they are very hard working and dedicated.

The students are of very high quality – they had to be to get admitted and can be dropped if their grades slip. They are an international bunch. My top student in my first tour was a girl from Uruguay. On my second tour, having already gotten her Master's and working in industry, she asked me for help in getting her into a Doctoral program in the USA. I made arrangement to get her a Graduate Assistantship at USC and she is now has her Doctorate and is a US Citizen working in the aerospace industry. The competition amongst the students is fierce, because the salary they will receive is a direct function of their Grade Point Average and faculty recommendations. But – in my opinion – they were not, during my tenures there, at least - as well prepared to work in industry as graduates from US schools – although they do have a better chance of making a big breakthrough later in life.

But I thoroughly enjoyed our many stays in Israel – an outstanding country for tourism (there are so many significant sites!). We have returned on several occasions to visit our relatives and to take our visiting friends on tour. I had later opportunities to be a visiting Professor the Naval Postgrad School in Monterey and at the Naval Academy, but they wanted me either for a whole year or for two consecutive quarters, and that was too much time away from home for us.

FINAL WORKING YEARS

THE START OF THE ISRAELI NAVY

My cousin Susan and her husband Dick first moved to Haifa in 1949. He, a WW2 US Navy communications officer –a Lieutenant - had been recruited to help start the Israeli Navy. He was to be in charge of their Communications – in a building situated on the Mediterranean side of the beach. They stayed for about 7 years; got homesick and returned to Michigan City, Indiana. After about 5 years, they yearned for Israel and returned to Haifa, where they have lived ever since - now on the 17th floor penthouse of a high rise overlooking the Harbor. In March, 2011, they sent me an email that featured an article that Dick had written many years ago. And THAT's the bulk of this story.

I only met cousin Susan once just before she went off to Israel with her new husband and baby. We met at her Uncle's place in Atlantic City. Her Uncle was my Mother's brother-in-law. She was a beautiful girl then – just as she is a handsome woman now. The reason I didn't know her was because she lived in downtown Philly and was in the JAP (Jewish American Princess) 'society' Dancing Class that was 1-2 years older than my Group. Dick, a Clevelander, was a student at the University of Pennsylvania where he met Susan. After graduation, he was a businessman before he first went to the old country. While he was there, he attended the TECHNION and got his Doctor's Degree there, and then became a Professor. He retired from teaching just about the time we arrived in 1989.

This is his story, taken from the CLEVELAND JEWISH NEWS PAPER, Saturday, March 12, 2001:

Battling the Japanese, injustice during WW II
By RICHARD D. ROSENBERG
Special to the CJN

CATCH A ROCKET PLANE

The year was 1945. The Okinawa invasion fleet was assembled in a huge lagoon in the Caroline Islands in the Pacific Ocean. I was a lieutenant on a destroyer called the Harding. One night half the officers on my ship were invited to a big barbeque on Mog Mog Island. We loaded our boats with steaks and beer, and with stewards along to do the cooking, went ashore.

There was nothing but sand and palm trees on Mog Mog. Some of the officers built a huge bonfire and the flames seemed to climb to the tops of the trees and up to the spangled sky. Just before we were all to return to our ship, an officer came out to lead the men in singing. The word was quickly spread that the officer was none other than Eddie Duchin, a famous bandleader who was now the captain of a destroyer escort. A little beery, and very much aware of where we were bound though not really conscious of our own possible fate, we began to sing to the tune of the "Wiffenpoof song":

"We're poor little sheep who have lost their way
Baah, baah, baah.

We're little black sheep who have gone astray,
Baah, baah, baah.

Gentlemen songsters off on a spree;
Damned from here to eternity
Lord have mercy on such as we.

Baah, baah, baah."

My best buddy Mal Cairns was standing by my side singing with me. Little did either of us know what was soon to be.

A few days later, on March 22, the Harding was ordered to sweep mines in the harbor of Naha, the principal city of the Okinawa islands group in the South China Sea. Despite having been sent on what could have been a suicide mission (with no air or large ship support), we met with no opposition whatsoever.

The invasion itself began on April 1. Again our forces met with little resistance, but at sea, the situation was quite different. On April 6, a massive Kamikaze attack was mounted against the U.S. fleet, and particularly against the Radar picket-ships that had been stationed in rings at some distance from the main island of the Okinawa group. The Harding was one of these, off the island of Aguni Shima. During this attack, some 240 Japanese planes were shot down. The Harding took a near miss that damaged the steering mechanism and we were sent to a nearby anchorage for repairs.

When we returned to our radar picket station on April 16, just three days after my 24th birthday, a group of planes attacked us in pairs from both the stern and the bow. We shot down the first seven attackers, but simply couldn't fire fast enough. Two planes approached; the first plane crashed into us at the water line alongside the Harding's forward ammunition storage. It carried a 250-pound bomb that exploded as it hit the ship. That touched off another explosion of about 1,000 five-inch shells.

I was in the Combat Information Center about 30 feet above where the plane hit. I remember not the sound of the explosion, but its impact, like some tremendous collision. On the deck examining myself soon after, I thought out loud, "Son of a bitch, you're O.K."

Running past the dead and wounded, I had to overcome my fears and go down to the wardroom, where I removed the secret codes and placed them in weighted bags to be jettisoned. By this time the water was up to my hips. A body floated by: my buddy Mal's.

The ship began to sink in a sort of dive, but the captain ordered us to reverse and the ship, although very low, began to right itself. We backed all the way to the Kerama Retto anchorage and were taken off the Harding to an attack transport being used as a temporary refuge for officers from the many damaged and sunken ships. And there the waiting began.

Landsman

I went up to the deck to reflect in solitude. The massive headache that followed the hit was beginning to subside. There was really nothing to do but listen for the next air raid alarm, and the alarms were almost constant.

As I sat there, another officer appeared. We exchanged routine questions and information about our ships, our "jobs," when we had been hit, etc. Then came more startling details. We were both from Cleveland, both communications officers, both Jews, and the officer's aunt Lillian had been my Hebrew teacher.

Needless to say, we became inseparable buddies, spending the "non-air-raid" time reminiscing and playing two-handed bridge. It got so we didn't even stop playing during the attacks, as there were no battle stations, no safe place to go to, nothing we could do anyway.

But I was kept busy with another task. The Harding was put into a floating drydock for repairs. Half the crew was sent back to the U.S., but I was one of the "lucky" guys chosen to remain. While we were anchored off Kerama Retto, I had a visit from a cook 2nd class in the dry dock crew. He was being court-martialed for refusing to obey an order and I had been recommended as his defense attorney. The man was a baker. One day, as he was drawing flour to bake bread from the ship's supplies, he discovered the flour was literally alive with weevils. When he saw the same was true of all the ship's flour, he went to the supply officer to complain. He was ordered to bake the bread regardless.

He refused and was court-martialed. The court, composed of officers from the dry dock, refused to permit me to present any evidence in support of his action or even mention the condition of the flour. So he was convicted.

In my closing remarks, I accused the court of bias and failing to perform its duty to rule justly. On the following Sunday, I was called to the dry dock to sign the protocol, which was to be forward to the reviewing authority. When I read the protocol, I found that it had been altered. My statement had been deleted, as had much other testimony. I was furious. I refused to sign. I told the court officers what I thought of them, and returned to the Harding.

I lay down to take a nap, only to be awakened because the Captain Ramage wanted to see me. He said: "What have you been doing with respect to that courtmartial?" When I told him, he showed me a letter he had just received from the captain of the dry dock, accusing me of "uttering scurrilous statements" about the officers of his command, and that he was recommending me for a General Court Martial (reserved for the most serious offenses).

Captain Don Ramage, an Annapolis grad, was the brother of a famous war hero who had been awarded the Congressional Medal of Honor. When I explained the situation further to him, he said: " Well, if they want to have some fun, let them try you; you may fry a little, but they will burn." Apparently he delivered this "threat" in person. As a result, and totally against Navy procedure, the protocol simply disappeared. The cook was never convicted and, outside of a very few anxious moments, I learned something about justice.

Sailing to Shul

On Shavuoth, I saw a radio message to all the ships at Okinawa announcing Jewish services would be held on the flagship anchored about 20 miles away. I contacted the few Jewish sailors, none of whom wanted to go. But I did. I asked the captain about having the ship's whaleboat take me ... which meant not having it and the two-man crew available for most of the day. No problem. There it was. The morning orders said: "Bring the motor whaleboat alongside at 0830 to take Lieut. Rosenberg to divine services." Incidentally, there was no rabbi in the area and a Presbyterian chaplain conducted the services.

Other little experiences helped pass this terrible time, because in addition to boredom, we were still being attacked by Kamikazes almost very day. Occasionally we would go to "the beach" – a small island where we could sunbathe, play baseball or cards, and drink the small allotment of whiskey and beer we were permitted.

One day on our way to the beach, I was showing the officer next to me pictures of my wife, daughter and my dad, who was in uniform. The guy asked me how long my father had been in the Navy and when I replied "about 30 years," an older officer not from my ship, interrupted.

"Your father's been in the Navy over 30 years? What's your name?" he asked. When I said Rosenberg, he said: "I know your father and I've even been to your home and met you." Small world.

Of course we were following the reports of the progress of the war in the Pacific and also what was happening now in Europe. It was clear we were winning, but we had no illusions about the Japanese willingness to die even when there was no hope of victory. So I was still not clear about my own future if the war didn't end soon. On August 1, we were ordered to proceed to Honolulu for further assignment. As we approached Saipan on August 6, we got the news of the bombing of Hiroshima. On August 8, we heard that Nagasaki had also been bombed and that Japan had surrendered.

I must confess my thoughts did not include any remorse for the bombings and the extent of the casualties and what it meant to the survivors and for us. I thought of only one thing: I have been spared.

I asked myself why? My closest friend on the ship had been killed. Why were he and the others killed while I was spared? I wrestled with this question many times and finally decided I could no longer consider my life just mine. Rather, I had been given this gift not only to (hopefully) return to my wonderful wife and little baby girl I only knew from pictures, but to do something with my life that was worthwhile for others. The real news of the Holocaust and the emptying of the camps had now surfaced, as were the reports of what was happening in Palestine. I saw this as a focus of my mission. Here's why.

Honoring a great man

My grandfather, Max David Stashower, was one of the first members of the Zionist Organization in Cleveland at a time when I believe there were only about 25 members.

Consequently, it was quite normal for me to espouse the goals of Zionism and the establishment of a Jewish homeland to absorb those who had survived the terrible tragedy of the Jewish people. I realized that the creation of a Jewish nation and my personal survival were connected.

I joined the Zionist Organization of America and became president of the Brigadier Kisch branch in Philadelphia. Through the ZOA, I made contact with Rechesh, the clandestine Israeli procurement agency of the fledgling nation of Israel. My tasks were collecting arms that Jewish soldiers, released from the U.S. armed forces, had brought home. I was also involved in interviewing and vetting volunteers for the Israel air force, many of whom were non-Jews. Soon I was given greater tasks, such as arranging for the financing and acquiring the machine tools needed to produce the bazooka.

One evening, I was approached by Teddy Kollek, head of Rechesh at that time. He asked me to go to Israel and become the head of communication and electronic warfare for the newly-formed Israel navy. I laughed at his request. "There were thousands of Jewish men in the U.S. Navy who are more qualified than I," I told him.

Here is a story only Jews could duplicate. He said, "If you take the number who saw as much combat as you, would you agree that's a smaller number? And if from that number you would take only those who had been communications, radar and sonar officers, it would be an even smaller number. Of those, take the number who are Jews, and from those who are Jews, take the number who are active Zionists. From those, take those who speak Hebrew. And from the ones who speak Hebrew, take the number who will go if we ask them. We know you're not an expert but you're all we have."

How does one turn down such an offer?

When I returned after the war and lived in Philadelphia, I was active in the Youth and Young Adult Division of the Allied Jewish Appeal. I even taught Sunday School.

But this is not the story of how I materialized my thanks for surviving the war. It's more the story of my grandfather's influence on my life. I have lived in Israel for almost 45 years. Max and Sarah Stashower have over 30 descendents who were born in Israel, who have served in the Israel Defense Forces, and who speak Hebrew as their first language.

And on the road to Jerusalem, there is a grove of 1,000 trees and a monument that says, "This grove was planted in honor of Max and Sarah Stashower by their grandchildren, and by their great grandchildren."

Richard D. Rosenberg, now nearing his 90th birthday, is a Cleveland Heights High School graduate living in Israel. His father, Abe, was an active Park Synagogue member after a naval career spanning both World Wars. His mother, Minnie Stashower Rosenberg, was the first American-born Hebrew teacher in Cleveland. (Dick died in Feb., 2012).

ABET ADVENTURES

In 2005, ABET formally changed its name from the "Accreditation Board for Engineering and Technology" to ABET, Inc.

This allows the organization to continue its activities under the name that represents leadership and quality in accreditation for the public while reflecting its broadening into additional areas of technical education

Are the new rules causing undergraduate engineering education to go to hell in a hand basket? ABET (Accreditation Board for Engineering and Technology) is the U.S. outfit that is responsible for ensuring that university Departments providing specialized (e.g., Mechanical, Aerospace, Electrical, etc.) Engineering and Engineering Technology degrees meet minimum established standards. The accreditation is accomplished by assigning teams of trained professors or, rarely, trained persons from industry who have had university teaching experience, to 'visit' an engineering college seeking accreditation or re-accreditation of its academic program. The college seeking this service pays all expenses accrued by the procedure except for the basic salary of the 'visitors'. Re-accreditation usually occurs on 6 year cycles unless some deficiency, applicable only to a specific Department in the Engineering College under scrutiny, is found. In this case, a shorter review cycle is recommended in order to give that Department time to correct the deficiency. ABET itself is a child of the various Technical Societies which support the specialized curricula. The Department being examined has the right to 'blackball' a Visitor, precluding prejudicial evaluations if there is some past history.

In the 70's, I was honored to be named a 'visitor' to Aerospace and Aeronautical Engineering programs by my technical society, the AIAA – the American Institute of Aeronautics and Astronautics. I was duly trained in the voodoo protocol entailed in a proper evaluation. I remained a candidate visitor for over 15 years, and in that time was selected for 5 evaluations. Obviously, to be, or to become, an 'unaccredited' Department is a blasphemy in the eyes of the Dean of Engineering, the University, the state legislature for state-supported institutions, and the Lord.

I was pretty much at peace applying the criteria that I had been taught in a special class that my technical society insisted their Visitors must take before their first assignment. The criteria were then aimed at determining whether the Department under examination was up to par in comparison with other accredited aerospace engineering programs throughout the country. The visitor spent about two and a half days working with the Dean of the College, the Department Chairman or Head, the Faculty, the Research and Teaching Assistants, the student president of the Student Branch of the AIAA, and some selected students usually picked at random. We looked over the vast 'book' that the Department had prepared detailing their academic program, including the required course of study, the Faculty work load and research programs, the number of students per class, typical exams, and student achievement statistics. I was always impressed with how uniformly good the engineering training seemed to at the undergraduate level. I soon figured out that the reason one college was said to be 'better' than the others stemmed strictly from their graduate program and the amount of contract money they were able to acquire from outside sources for research. So, 'visiting' and making a correct accreditation finding was not difficult, and was interesting and fun. More important, I thought it was being done right!

Somewhere in the '90s, things apparently started changing. Engineering college budgets were being cut. The 'fat' had to be drained from the undergraduate academic program and acquisition of research to support faculty became more and more important. I suspect the Engineering Deans supplicated to ABET to give them more slack to operate with reduced assets. But, from my latter year observations, I fear that the net result has been to lower the standards under the ploy of 'giving the customers just what they asked for'. I uncovered this plot during one of my periodic visits back to Iowa State and a discussion with a tried and true retired faculty member who still was putting out his "Rag' – the official Department monthly newsletter for students, faculty, alumni and broken-down ex-Department Heads. He told me a shocking story: Due to lack of faculty, some basic engineering courses were being taught to classes with more than 80 students! The reason this was shocking was because one of the 'visiting' rules that I had lived by was to start 'dinging' as soon as basic classes got over 30, at the max, students.

I couldn't believe what I had heard, but soon confirmed it with several of my former faculty members. I rushed to the then brand new Department Head and to the Dean asking for an explanation and advising them that they would never get their program accredited! They both came back with the new party line: "Don't fret, the new ABET criteria are based now on 'OUTCOMES' – not how you get there. We now rely heavily on the feedback we get from our 'customers' – the aerospace industry that hires our graduates – to tell us if our education is up to snuff and what they want". "Oh", I said thinking that I was now in Wonderland with Alice, "Thank you for straightening me out". I went home and stewed for a few days – then took action:

September 3, 2003

Larry Nixon, President
ABET
111 Market Place, Suite 1050
Baltimore, MD 21202-4012

Dear Larry:

I am a former ABET Aerospace Engineering Accreditation Visitor who operated for many years and many visits under the 'old' criteria. I have just returned from a visit to the College of Engineering of a major state-supported University whose name I shall not mention. I was appalled by an undergraduate classroom size situation there that I was told would now be acceptable under the new 'goal oriented' ('outcomes') criteria, with which I am only partially familiar.

In my time, if a Department showed several class populations for required technical courses not of the "Physics 101" lecture type over, say, 30 students, they would be subject to criticism and a 3-year recommendation. Here, I found several such classes up to a remarkable 83! I was told by a Faculty member, that if such overpopulation fits in with the stated departmental goal scheme, so be it! "This can't be true", I said to myself.

I would appreciate you, or someone in your organization, telling me that I got the wrong information or, if this is true, that ABET will review what I believe is a great step backwards, made to accommodate schools suffering from budgetary woes.

Thank you for your attention.

Dr. RF Brodsky
Professor (Ret.) of Aerospace and Mechanical Engineering
University of Southern California
110 the Village, #410
Redondo Beach, CA 90277

Not receiving an answer to this perfectly civil inquiry, I moved on. I had just read that an old friend, Dr./Prof. Leroy 'Skip' Fletcher, ex of Texas A&M and now head of engineering at NASA Ames Research Center, had just been elevated to Honorary Fellow of our technical society and been elected a Director of ABET. So-

Subj: What's with ABET?
Date: 12/8/03
To: Sfletcher@mail.arc.nasa.gov
CC: (mikeg@usc.edu), (melsa@engr.iastate.edu)
CC: (pjay@iastate.edu)

Skip:
   Congrats on HF!!! Now - I have another bone to pick:
   With your new ABET hat on, could you try to get at the bottom of why ABET has not acknowledged my deeply concerned letter (below); first sent to the President, and later to the Exec. Sect'y. I am truly worried that ABET has sold out and that the quality of engineering education has dropped a peg.
            Best regards.   Bob

He answered promptly, ironically beginning an Iowa State horse opera. For starters, let me tell you that Dr. Dave Holger, who, several Chairpersons later, had succeeded me in running the Aero. E. Department at Iowa State and was now an Associate Dean:

Subj: ABET
Date: 12/22/03 3:48:03 PM Pacific Standard Time
From: sfletcher@mail.arc.nasa.gov (Skip Fletcher)
To: Rfoxbro@aol.com

Bob,

The proper person to send your letter to is David Holger (holger@-iastate.edu). He is Chair of the EAC (*ABET's Educational Activities Committee*), a former AIAA Visitor, and one of the former AIAA Reps to the Commission. So he is the one to answer your question.

Skip

In answer to my query, Dave wrote:

Subj: RE: What goes around, comes around
Date: 1/2/04 2:47:57 PM Pacific Standard Time
From:holger@iastate.edu (Holger, David)
To: Rfoxbro@aol.com

Bob,

I don't have an explanation of why Larry or others have not replied to your e-mail. I can however provide a bit of information that may or may not explain the EAC's position in terms of the current (EC 2000) accreditation criteria.

The bottom line philosophy of the current engineering accreditation criteria of ABET is that programs must demonstrate that they are preparing graduates for successful careers in engineering. It is a results oriented approach rather than an approach that specifies how the results are to be achieved. While I agree that an engineering program with only large lecture classes is unlikely to do an acceptable job of preparing graduates for engineering careers, it is certainly possible that an appropriate mix of large and small classes could work. If the students are learning the things they need to learn in a large class, then that class is probably doing a good job of preparing them in the topics that are covered.

It is not true that a department/program can establish their own goals without regard to the expected preparation of graduates in the discipline. It is true that different programs can choose a different mix of educational mechanisms and courses to prepare qualified graduates. Program evaluators are expected to examine the evidence presented by a program and to decide if the results are adequate – that is are graduates of the program prepared to become successful aerospace engineers. Do they know and are they able to do the things that industry expects of successful engineers.

I am not going to argue that all the details of how this philosophy has been applied have been perfect in all cases. However, I will argue strongly that a results oriented approach is better than a requirements oriented approach. The expected disciplinary/sub-disciplinary coverage has not changed from the "old" criteria. If anything the expectations are higher now than ever.

Dave

David K. Holger
Professor of Aerospace Engineering
Associate Dean for Academic Programs and Budget
Editor-in-Chief, Noise Control Engineering Journal
College of Engineering
104 Marston Hall
Iowa State University
Ames, IA 50011

This represented the final thought on the matter – or so I suspected. I sent it around to several other 'old criteria' visitor friends, all of whom agreed with my finding that engineering education had taken a step backward. While throwing up my hands in exasperation and frustration, I still can get in two last licks:

The first was engendered by the recent arrival of a new issue ABET magazine, which announced that old Iowa State chemical engineering professor friend Dick Seagrave was to be the upcoming ABET president in 2005. Attaching most of the past correspondence, I sent him the following:

Subj: ABET to Seagrave
Date: 4/12/04
To: seagrave@iastate.edu

File: C:\My Documents\ABET #1.doc (20480 bytes)
DL Time (32000 bps): < 1 minute

Dick:
Congrats on your forthcoming ABET gig. As you will see from the attached (and the conversations attached below with Skip and Dave), I have a nit to pick with the Outcomes criteria. I hope you will foment some conversation in future Board meetings. No matter what any criteria say, 50 and above students in a basic engineering class is thumbing your nose at the system.
Regards and good luck in your endeavors. Bob

I did not get a reply, but recalled another past ABET-associated incident to grouse about:

I was only trying to be helpful when I replied, on December 20, 1993, to an earlier letter from Dr. William B Streett, Dean of the College of Engineering of Cornell University. His letter turned out to be a paean to the glories of Cornell engineering, and in no way indicated that he knew that I was an alumnus of his College, nor that I was an academic colleague. My letter did not disagree with his appraisal of the superiority of his school, but I did carefully take advantage of his inquiry to note my high state of dudgeon:

"Dear Dean Streett:

You recently sent me a letter which began, 'Dear <u>Mr.</u> Brodsky: You are one of a select group of engineers periodically surveyed------- Your opinion of our College, then, is important to them (i.e., potential students) and to us.' I suspect that one of the reasons that I was among the chosen few is because I am an ABET visitor.

Despite your personal lack of insight / foresight, which I will develop in a diatribe below, I still think <u>Cornell engineering is outstanding</u>, and recognize my personal debt to it, but---

- It is true that I am <u>Dr./ Professor</u> (not Mr.) RF Brodsky of the University of Southern California
- That I was awarded the AIAA/ASEE outstanding Aerospace Educator honor, and am a Fellow of AIAA
- That I have been listed in "Who's Who in America" since I appeared in TIME magazine (Feb. 1, 1963) as inventor of a space lifeboat
- That I wrote the description of "Aerospace Engineering" in the latest College Board guide
- That I wrote to you personally about two years ago noting that, in spite of my deep friendship with my Cornell classmate and your faculty member, the right honorable Franklin B. Moore, Cornell engineering had refused to enter the 20th century by ignoring the teaching of things astronautique – and was subsequently patted on my head
- That I was head of the Aero. E. Department at Iowa State University during the 70's, and
- That I am (was) a Landsman, being BME Sibley School, class of '46

FINAL WORKING YEARS

My quest for a little 'quid pro quo' was, as usual, unheeded, unheralded, and unrecognized. Even worse, to the best of my knowledge, Cornell still does not teach "space" in its engineering program. But then, neither does Cal Tech, despite my similar preaching there.

POWER POLITICS AND FUSION

I wrote this rant over 11 years ago, and was struck by the lack of real progress and energy insight that has NOT occurred in the interim. In year 2011, gasoline prices in the United States were near record highs hovering around $4.00/gallon – just as it was when I wrote this piece. Natural gas is now being looked at and may be a stop-gap as long as the supply lasts. But, in reality, non-replenishable coal is still being gouged from the ground and polluting the atmosphere; equally polluting oil is being sought in ever-more dangerous (to the environment) places; and now, the harvesting of shale promises new sources of oil and more challenges to the environment. True, we have produced agriculturally-grown fuel at great expense and have utilized 'clean' solar and wind power. But surely these are recognized as band-aid solutions that can not replace 'dirty' oil and coal as major sources of energy. And recent developments in Japan underline the futility of building more nuclear fission plants. As I have been preaching for over 30 years, the Hydrogen Economy, abetted by ('clean', if you can contain the neutrons) fusion reactors seems to be the only logical and affordable way to save the earth in the near (next 100 years, at least) term.

My record of saving my rants and ravings isn't perfect. A point in question is that I can not produce the April or May

of 1979 letter that I must have sent to Senator Ted Kennedy, one of my favorite then-living politicos. The reason for its disappearance must be that it was written while I was on sabbatical in Hermosa Beach, and got lost when we returned to Iowa. As I recall, the outrage that I expressed in my letter to him was a result of the disastrous gasoline shortage and price gouging that had taken place as the summer vacation period was coming on. As I write this in August 2000, the same jump increase in gas prices has been extant for almost 6 months, the cost of electricity in the San Diego area has gone through the ceiling, and the government has again seen fit to do nothing about it.

I think it may be edifying to quote some of Teddy's thoughtful return letter to me of May 22, 1979. I think you will better understand what some of the problems were then, and will be able to extrapolate that, twenty years later, nothing has changed:

"Dear Dr. Brodsky:

Thank you for expressing to me your views on national energy policy. I share your concern that rising energy costs are making it harder to balance family budgets. Home heating oil prices have soared in recent months, and the cost of gasoline is not far behind.

I believe that President Carter's decision to end controls on U.S. oil prices is a serious mistake. ------ The result is that a family of four will pay an additional $300 a year for energy because of decontrol, yet decontrol will not provide any significant savings in oil imports, either through increased conservation by consumers or through increased production by oil companies.

FINAL WORKING YEARS

Although free competition is ordinarily the best regulator of prices, the OPEC cartel makes a free market impossible at this time. Therefore, controls should be preserved. Our own government, not a foreign cartel, should make the decisions about what price levels are appropriate for oil produced in the United States.

The Administration has also proposed a windfall profits tax to recover part of the excess profits the oil companies would gain if oil is decontrolled. I believe they should not be allowed to receive the windfalls in the first place. But the windfall profits tax that the Administration has proposed is very weak. It would recover only about 10% of the windfall that will be produced by decontrol. -----------

In the longer term, our energy policy must be based on three themes -- diversification, productivity and competition.
First, we must diversify the sources of our oil, so that we become less vulnerable to the actions of the cartel in the Middle East. For instance, both Canada and Mexico have very large supplies of oil and natural gas. --------
Second, we must increase our productivity of our energy use in our homes, factories, and commercial buildings. ---------
Finally, we must increase competition, both from new sources of energy and within the oil industry. -------

-------------. I appreciate your interest in contacting me on this issue of such importance to our country's future."
    Sen. Ted Kennedy

The issues cited by Senator Kennedy remain the same today – yet a far-reaching solution is nowhere is sight. The Uber-Politics of oil supply and usage appear to be beyond

the grasp of politicians, but not beyond the ken of scientists / engineers. An alliance between the two; the same type of alliance between the good folks that gave you the Atomic Bomb and the Moon landing, can and should establish a 20-30 year program to relieve the U.S. from total dependence on the oil economy!

A brave, far-sighted politician, akin in guts to the Senator's brother, Jack, should propose a plan to move us into the <u>Hydrogen Economy</u>. The hydrogen – all that one can possibly need forever – would come from safe nuclear fusion boilers turning sea water into Hydrogen and Oxygen. When these elements are ignited, they produce more energy than the equivalent amount of gasoline and produce only uncontaminated steam as a by-product of their combustion. As a result, the environment is also saved and, in the United States at least, the smog problem disappears and the world's ozone layer depletion problem is greatly alleviated. True, the investment will be considerable, but, once operational, the pay back will be quick and permanent.

We are so close to being able to maintain and contain a controlled fusion process, that only an extra '5 bucks' and guaranteed long term support at this higher level, is all that is required to put the scientists over the top. Of course, the implications of such a change on world balance of power and abandonment and replacement of many of the normal fixtures modern life are humongous. But, the funny thing is, the minute a high government official announces that such a program will be proposed, and people are made to believe that we have the will to take such a step, the price of oil will miraculously drop like a stone. Then, if we want to, we can postpone the inevitable decision to take this momentous step for many years in the future.

In the past, I have sent my precious advice to Senator Kennedy on other earth-shattering occasions, such as the

Iran hostage crisis. So far, he has never followed up on my suggestions. I am not holding my breath in this instance, although I did send him a copy of this latest advisory. However, many years later, the subject reappeared in the form of a "Commentary" article in the Op-Ed page of the September 2, 2002 *LA TIMES*: In the article, *"The Power To Change The World"*, written by author Jeremy Rifkin, the key point was "A hydrogen-based system replacing our reliance on oil would revolutionize society". I felt Rifkin did not go far enough into realities and submitted the following (not printed):

Letter to the Editor, LA Times
September 2, 2002
From: Prof. (USC) Robert F. Brodsky, Doctor of Science (1950)
   Redondo Beach, CA 90277

THE HYDROGEN ECONOMY

In his 'Commentary' article entitled, "The Power To Change The World" in today's paper, Jeremy Rifkin rightly pointed out the world shaking implications of going on to the 'Hydrogen Economy' – but he completely missed the two other major implications of attempting to move in this direction.

First of all – where does an endless supply of inexpensive hydrogen come from? Obviously, from the ocean, of which there is no dearth. But, for the hydrogen economy to be real, competitive, and affordable, ocean water must be dissociated into its elements – hydrogen and oxygen – by, and only by, nuclear fusion methods. Any other separation system now known is simply not sufficiently efficient or affordable or, as in fission, is not socially acceptable. Thus, some brave highly placed person (more on this next) in our government, in the manner of John Kennedy's bold proclamation to 'go to the moon in ten years', must move for the formation of a new Manhattan Project – dedicated to take all infrastructure steps necessary to move us into the Hydrogen Economy in, say, 25 years. The key of course, is to throw significant money into long neglected fusion system development, whose progress in the last ten years has been negligible for lack of such funds and intent. This has not been done for political reasons!

This brings us to the real problem why "Hydrogen Economy" proponents have been so silent. Politically, the uphill battle seems insuperable. Our politicians are owned or belong to the international oil cartel. This has always been the case and, alas, may always be so in he foreseeable future. From our President (i.e. George Bush) *and many of his Cabinet on down to our Congresspersons, most of who delighted in Enron's success, oil and its politics are their life blood. Who among them is foresighted enough to take, or even breathe, the hydrogen step? None, I fear –ever!*

The United States, its allies, and its oil-producing 'friends' all want to maintain the status-quo – even though political accommodations are now being stretched almost to the breaking point. They will find the compromises necessary to continue the blackmail and terrorism to the detriment of improving the world's energy and environmental conditions.

The funny thing is that the United States, by merely stating that its goal is to move onto the Hydrogen Economy in 25 years and putting a few more bucks into fusion research to show our sincerity towards this goal, could cause significant changes into everyone's thinking. The oil blackmail would stop in its tracks, and reasonable people would begin to talk.

I sent a copy of this piece to several friends to show them what a seer I was. One quickly pointed out that I had absolutely and forever ruined the point of the 'Letter' by using, in the now-corrected paragraph 2 above, the word 'fission' instead of 'fusion'! I tried to correct the mistake with a quick 'oops!' e-mail to "Letters to the Editor' Department, but I imagine it was too late to have an impact. What a dummkopf!

RETIRING WITH RELUCTANCE AND A LAW SUIT

In the late 80s, the Aerospace industry was 'enjoying' one of its periodic downturns. Layoffs were required, and it was human nature for the industry to seek out candidate from the 'old' and 'expensive' categories. At age 63, I was not quite mentally prepared for retirement, always having planned on getting closer to the traditional 65 before hanging them up. Moreover, in another year, I would have 10 years with the Company – and this milepost would increase my retirement pension considerably. Also, I did not really know if I could afford to give up wage-slaving.

My lay-off notice, when it finally came in a memo dated April 22, 1988, was not a complete surprise. I had been warned and I empathized with the company plan to relieve their payroll of the older high salary engineers in this critical post-cold-war time of aerospace industry belt-tightening. But, because it was the first and only lay-off notice that I had ever received in a 48 year career in the 'Biz', it still came as a shock. In view of my seniority and international recognition, I really didn't think that they had either the chutzpah or the cojones to make such a move which, in today's spin-infested society, is labeled 'downsizing'.

My job in the last couple of years had been to act as a systems engineer in short term advanced studies or to help write proposals for new work. I had no permanent home or sponsor. A few weeks before the fatal notice, my Boss had suggested that I look for an essential spot in a funded project; fully knowing, I think, that such assignments were all being filled, post haste, by the younger less experienced system engineers that reported to him. I recognized, however, that he was giving me a 'heads up' to protect my ass and his. When he gave this advice, I looked at him like he was crazy. "Dick", I said, "You must be kidding! I thought I was unfireable." He only smiled sadly and told me to pay attention. I responded to the new situation thrust upon me by taking four essential actions:

First, before receiving the formal notice, I wrote and distributed an internal memo sent to my friends in the Company advising them of my availability and desire to find a home where I could weather the storm, while also unfruitfully applying for advertised position openings within the Company.

Second, still before the notice, I took on a Financial Advisor – a former TRW engineer with a Doctor's degree who had become fascinated by finance and who had been recommended to me by a trusted member of my 'Old Farts' sailing crew.

Next, but after the death-knell notice, I successfully bought more 'adjustment' time by getting an extension in my termination date due my impending stint on jury duty.

Finally, availing myself the recently passed Congressional bill which permitted people to see their own company files, I requested, poured over, and made critical copies of information I deemed might be pertinent should I decide to initiate a grievance.

A few fact-filled consultations with my new financial advisor immediately relieved the pressure. It became eminently obvious that I could retire comfortably without reducing my standard of living, and would not need to seek work. So I relaxed my internal brown-nosing job-seeking efforts and, cheered by some very flattering unsolicited letters of praise I had found in and copied from my personnel folder, began thinking of fighting back. My new attitude of insouciance and goal-expansion was further brightened by an offer from a friend and former colleague to join his burgeoning start-up company as a consultant, working as much as I desired.

It soon became obvious to me in my efforts to find a new home in the company that the "fix" was in! Upper Management thwarted any effort by my friends in the Company to fill their appropriate openings. Those jobs were going to my compatriot younger, less well paid engineers who they wished to stockpile during the downturn. Not that my fellow workers were not competent, but they were certainly less qualified. After all, I was also a professor at USC and had taught the niceties of spacecraft design to literally hundreds of graduate students.

I applied for two advertised high-level 'in–house' positions both of which carried bonus pay features. One of the job descriptions looked as if it had been written with me in mind! In both cases, I was refused by a curt "experience not applicable" notation, which I thought at least showed that the company management had a healthy sense of humor. I checked out quietly on July 1, 1988 to face an uncertain afterlife.

As I later thought about it, I came to believe that I had a prima facie "Age Discrimination" complaint, which I decided to exploit as a hobby following my termination. Money was not my object, though I did wish to make up for the future pension income I would lose from several sources by not working full time 'til age 65 and 10 years. Also, probably hurt in the ego more than I wanted to admit, I decided I wanted to show 'them' that they could not push around working stiffs of my caliber. I came up with a number of $500,000 in damages consisting of one - plus years of lost salary and consequent lost pension incomes for the next ten years. I visited with two law firms who specialized in age discrimination cases, and who took hopeful cases on contingency. Both assured me that I had a good case, but that I should expect to invest about $10,000 of my own money to pay for depositions, etc. But, they said, a large settlement would not be a surprising outcome.

Not feeling that frisky, since I was just groping my way into retirement, I turned to the State Of California *Department of Fair Employment and Housing*'s downtown LA office to seek assistance in my quest. The DFEH did not charge for their services if they accepted the case, but warned that their settlements were limited by law to considerably less than the half mill that I had hoped for. But, I would only risk my time, plus mailing and commute expenses if I opted to have them take on my case. Their legal expert listened to my story and opined that I had a good claim and that they would be willing to work with me. He then suggested that I write up a bill of particulars to assist them proceed. I came up with a long diatribe whose main features I will excerpt to give a flavor of my bitch:

"11/28/88

MY CASE FOR AGE DISCRIMINATION

DR RF BRODSKY

BACKROUND

At age 63 and earning ~$80,000 (plus Bonus) per year from the TRW Space and Technology Group at a time when I was at the peak of my technical prowess, I was unceremoniously "laid off ", less than two years short of achieving full (10 year) pension vesting, I was told that all possible positions had been surveyed for openings and/or "bumping" possibilities, but that "no weak sisters were found, and it would be difficult to 'bump' lay off persons doing even an average or slightly below average job". Moreover, "jobs are now controlled by the Program Offices. All top jobs are now taken (see later comments about "squirreling away" favorites) and they just want to retain low paid technical people". I was "looked on as a future tools developer and no loose research funds were available".

At the time of my lay off, my credentials were impeccable. I was an internationally known engineer and educator (then, Adjunct Professor of Aerospace Engineering at the University of Southern California) with over 40 years experience in the aerospace industry. I am listed in "Who's Who in America" and several lesser national and international Who's Who's. I am a world wide lecturer on space and educational subjects, and likewise am well published in US and foreign technical journals. At the time of my layoff, I was one of four active engineers (of the others, two are Vice-Presidents and one is the company CEO) who have been awarded "Fellow" status in the technical society (the AIAA) that serves the aerospace industry. Additionally, I am a Fellow in the LA-based "Institute for the Advancement of Engineering", an honor I received in 1971. Two company VP's just recently were so-named, making a total of three, to my best knowledge. I am a Professional Engineer in the states of California and Iowa, a university academic program accreditor, and an Advisor at various times to NASA, the Air Force, and the National Academy of Engineering

MY LEGAL CASE

I believe I can establish a prima facie case of age discrimination as the basis for my layoff, for the following reasons:

- I was discriminated against in the lay-off selection process (approximately 5-6 people in my immediate organization of ~50 people) because I was within two years of full retirement eligibility; my advanced age (despite a near perfect health record); my substantial outside commitments which brought favorable publicity, but no revenue to the company; and doing research & consulting; thus not damaging an on-going program by my dismissal.
- It is my further belief that they established a pattern of age/salary discrimination by the selection of others that they laid off in the period surrounding my notice. I have made a list of others that I know were laid off; and all appear to be in their 50's and 60's, and relatively high paid-------
- During this same period (a sharp business decline was easy to forecast) my management took the opportunity to 'squirrel away' many younger presumably lower paid favorite-son engineers (I believe professional jealousy may have played a part against me) into safe haven jobs in Program Offices that had adequate long term funding, or into research work of a longer term nature-----
- The most damaging action (fully documented!) taken by the company was the incomprehensible refusal to consider me for a job for which I was superbly qualified (and for which I applied shortly after receiving my notice) and which would have enabled me to again be placed on bonus (more about this later) status. The turn-down cited "minimum requirements not met" and "inappropriate experience/skills". Nothing could be further from the obvious truth!
- The timing of my lay-off coincided with my withdrawal, at management's strong suggestion, from the various outside activities with my technical society which had placed the company name in the forefront on the local technical scene, and for which I had a separate overhead budget. -------

In fact, my treatment vis-a-vis the bonus situation has always been a bone of contention, ---- I have documents to show that I was lured away (note; this is an exaggeration!) from the campus and a tenured-for-life position with the promise of much gold via the bonus process. I was not informed that the bonus process could be turned on and off like a faucet: I was dropped from the bonus rolls as soon as it was decently feasible! I re-achieved bonus status due to a later promotion. I was relieved of this job when the position was abolished after I told management that their expectations were not, in my opinion, realizable.

Finally, note that during the 8+ years I was employed at TRW I always received an annual raise in addition to the substantial bonuses I received several years. In fact, I have available several recent voluntary letters of commendation about my work. ------"

On December 7 (a day that will live in infamy!), 1988, the DFEH informed me that they had started a file on my case and that I must give them a timely notice of my intent, should I choose to do so at any future time, to initiate a private lawsuit. Thus began a series of meetings and information exchanges in which I named names and provided various witness letters and letters of commendation from my company files. They, in turn, sought information through the offices of the company's internal lawyers, who – from remarks made by my project officer – stonewalled as best they could, while gladly providing statements from my former colleagues as to my shabby performance; quite contradictory to the nice letters of commendation that I had received (and had copied from my file) from several of the very same people! In some cases, the slander was downright nasty, but I recognized that my old friends were under the gun, and that all of this was merely 'show biz'.

On July 6, 1989 – a year after my lay off – by the command of the Director of the DFEH – a formal 'accusation' notice was served on the senior counsel of the company. This included a notice of hearing, the accusation, a copy of my complaint, a notice of defense, statement to respondent, right to interpreter, administrative procedure act and 'new regulation' – i.e., all the boiler plate documentation needed to legally support the righteousness of their suit. The DFEH further established tentative dates for a hearing for September 18-19, and provided company counsel with papers to request such a hearing.

No such hearing proved necessary. Apparently the company, at the advice of their counsel, decided to fold and sought a settlement. I signed the agreement on July 28, 1989, accepting recompense, in two bundles, for back pay as well as for 'emotional distress'. In total, the agreement yielded a bit less than half a year's salary – but the 'win' was very satisfying. As part of the agreement, I had to agree to not sue the company in any other related action, and not to disclose the settlement financial terms. The company, for its part, had to agree to take steps to that this unfair practice was eliminated and subject themselves to a future review to determine that they had complied. The same day, I sent out a letter to the friends who had supported me:

"Dear Friends:

You will be pleased to learn that on this date I signed an agreement negotiated between the black hats and the white hats (the State of Calif. Department of Fair Employment and Housing) which will bring me modest recompense for the shabby way I was treated (and thus further ease my so far successful retirement! Alas, I will have to pay income tax on half of the settlement; the other half being for my "pain and suffering").

I could have chosen to take them to court and might have ended up with a real bundle of moolah, but I was advised that this would probably take another year; mean an $8 - 10,000 investment in deposition and other costs with no guarantee of recovery; and a 40-60% split with the lawyers (my 60%). So, since we'll be leaving for Israel in late September, I decided to take the money and run.

In letting them off the hook, the company was admonished to a) post notices advising employees of their rights and how to redress grievances, b) use non-discriminatory practices when selecting employees for lay-off, c) review their policy on ranking of employees (which is purely subjective and " what have you done for me lately" now), in addition to my pay off.

So, I guess I'm vindicated and reasonably happy. I will forever of course wonder if I shouldn't have gone for the big banana -------- Anyway- the point of this note is to thank each and every one of you (and, my wife, dog, and heirs join in) for both your moral and verbal/written support for this great cause against the sinister military-industrial complex that has nurtured us all. For you, my everlasting gratitude and a sailboat ride whenever you're in the neighborhood. We will return from Israel in mid-February. Have a great summer, and for real fun - sue somebody!"

For all my ventures into the mysterious world of the LAW, this of course has been the only one with a big money pay-off. Still, as I later thought about it, look how smart, albeit somewhat unethical, the Company really was! At very little expense, they simultaneously got rid of a high salaried engineer past his peak; saved two years of paying his high salary; saved on payments of his pension, which would have been greater at attaining of 10 years of service. They actually came out of it smelling like a rose, with only a slap on the wrist – for by the time the DFEH stomped down on their evil practices - they had achieved the 'mean and lean' watered down complement of engineers that was right for the time. Although word of my victory got around and I received several inquiries as how to proceed, I am not aware of anyone else 'winning' a jackpot. I think my case may have been unique.

Even if I had had the guts to go for the big money and had won, as was probable, they still wouldn't have been hurt too badly. That's why they are a great company. My hat remains off to them. I bear them no ill-will. In fact, I remain today an active member of their Retiree's Association, and support their docent and travel programs. But, the beat goes on! On Sunday, July 1, 2001, many newspapers in the U.S. carried a story by Adam Geller, Associated Press Business Writer. Its headline was, "FORCED OUT AT 50? Older Workers Allege Age Discrimination in Job Cuts". An accompanying graph inset was titled, "Age discrimination complaints increased 15.4% last year".

Chapter 8

AIN'T RETIREMENT GRAND?

- EXPERT WITNESSING AND CONSULTING
- A MEDIA DARLING
- MY LOVE-HATE RELATIONSHIP WITH 'MACHINES'
- AROUND THE WORLD – SPRING '99
- THE 70th REUNION

Since retiring from the aerospace industry (TRW, 1988) and academia (Prof. of Astronautics and Space Technology, Univ. of Southern Calif., 1996), I have been writing non-fiction books; having started on my fifth in October, 2009. All my books, except *"Songs ---"* can and will be found on Amazon.com under the name of Robert F. Brodsky.

My first book (2006) was *"On the Cutting Edge"* – 'Tales of a Cold War Engineer at the Dawn of the Nuclear, Guided Missile, Computer, and Space Ages'. My second (2008), *"Songs My Mother Never Sang to Me"* is a compendium of the words of songs learned and sung over a lifetime; many from songfests at fraternity houses and the Ithaca Hotel. The third (2009), *"A Pilgrim Muddles Through"* consists of Annals from the worlds of Merchandizing, Morality, Authorship, Law, Writer's Workshop, an Outraged Citizen, Sailing, and Retirement. *"The World in a Jug"* came out in June, 2010. It is concerned with traditional jazz - New Orleans and Dixieland genre - which I used to play and now lecture about in my "Serenades for Mouldy Figges)".

This book, my fifth, *"Catch a Rocket Plane"* consists of memoir-like stories from my experiences as - an atomic bomb designer, - a space pioneer, - an award-winning

professor (Iowa State U, USC), - a globe-circling traveler, and - a world-class professional RETIREE - that were not covered in "Cutting Edge".

EXPERT WITNESSING AND CONSULTING

In the parlance of the streets, an 'expert' is "someone who's from outa' town". However, in legal circles 'Experts', by definition, are people who are supposed to know something in a specialized field much better than an educated generalist who is familiar with the field. But from my new-found experience, I believe that experts only need to know more than the layman, even if the layman is highly trained, as long as he or she speaks with sincerity and authority. But in the real rough and tumble world, an 'expert' is the best person you can find who is immediately available and who has reasonable credentials and doesn't charge too much. Thus, it is that I have been called upon on four occasions, the first two quite inadvertently, to act in the role of an 'expert witness' – a small fish in a pool of sharks.

My first two jousts with the courts came during my student and working days. Both were disasters – in the first I was abased and ridiculed and in the second, I was fired by my Attorney-hirer because I advised him to 'settle' ASAP. Both of these encounters have been described in my first book, "*On the Cutting Edge*". In this Chapter of "*Rocket Plane*", I cover three latter day outings into the consulting arena. In the first, I would have liked to describe in detail a case of import in which my side won for the nonce, apparently, due in part to my frank but unapt handling of the opponent lawyer who was deposing me. However, the crux of the matter is that I have been advised by a Lawyer that a hired 'expert witness' or paid consultant is apparently forever muzzled from disclosing anything pertinent about the cases in which he or she participated. Thus, in this instance, I can only comment on what it is like to be an expert witness. A second case describes the tribulations, working in a PRO BONO expert mode, dealing with the design of a novel piece of sports equipment and potential but now unpursued legal action seeking redress. The last story covers an interesting consulting experience in which I earned a lot of money, but little else – and whose outcome is unknown.

HI-DIDLY-DEE, THE EXPERT WITNESS LIFE FOR ME!

In the early 2000's, I was referred to a patent law firm who thought my credentials and experience were impressive enough to act as an expert witness in a hi-tech case they had taken on. What helped was that I had a Doctor's Degree in Engineering; was a 'certified' Professional Engineer in the great States of California and Iowa, and was a university-level Full Professor (Iowa State and USC) and had retired after 36 years of experience in the Aerospace Industry. While I can't relate the circumstances of the case, which is still – in 2011 – ricocheting around the courts, I can tell you about the experience of high stakes 'expert witness-ing', and so I shall:

I got the call from out of the clear blue sky. We talked for about an hour on various aspects of the case, and discussed compensation. "How much do you charge?" I had no idea how to answer this, and soon made my first big mistake. I churned over the fact that at last look, over 10 years ago, I got a thousand a day for my consulting services. This clocks out at $125 per hour, so I daringly said, "How about $150 per hour plus expenses", fully expecting him to scoff at this highway robbery demand. I dared to be so bold because I really didn't need the money, being now happily retired. However, he quickly said "That's fine!" and went on to other business. I later found out that no one would have batted an eye if I had asked for $300 per hour and more, say $400 or $500, during anticipated depositions and court appearances. Caveat Emptor!

During my part of the case, which lasted through the Summer and into the Fall, I learned to better appreciate lawyers and the job they do; changing my former opinion that they were a rather low form of humanity. The two ace Lawyers I was associated with, and their associates, quickly showed me how intelligent and how well trained, in both law and engineering, they were. And, of the importance of every nuance of language either to make things crystal clear or to weasel. And, of the importance of using the phone, rather than FAX or e-mail to express opinions – since the latter two had to be made available to the opposition. And, finally, of the importance of thinking before speaking – a lesson that has always come hard to me. Most of all, I found that they were very hard working and, in getting ready for appearances before the examining Judge the next day for example, thought nothing of working through a 20 hour day. I found myself being awakened in the middle of the night by telephone calls; sending out FAXes at all hours; and rushing to offices for conferences on demand -–and all for a bill and a half per hour!

I had to make many learned statements, all of which they rewrote in obfuscatory lawyer-ese. I had to draw sketches and make long-winded explanations of my position. I had a lot of fun and became more and more certain that my stance was correct as I read the pontifications of the other side's experts. So, when deposition time came, I was cocky and confident. I did not expect the other side's lawyer to be so sharp nor did I expect a 4-hour non-stop session. He insidiously probed into all aspects of my supposed expertise and ended up knowing my life history. I had been instructed to be wary and circumspect and that, for the most part, I was.

However, when he asked me to give my detailed opinion of his expert witness' written testimony about the key point of the argument, I just couldn't restrain myself – it slipped out quickly, "He's full of shit", I proclaimed. The ensuing silence was deafening, and afterwards my sponsor was quite perturbed by my un-professionalism. However, I honestly think that when the Judge Examiner read my testimony he must have at least smiled, and at best swung his opinion to our side.

You would think that it was a happy ending, no? Well, when Income Tax time came around, about half of the unanticipated income had to be returned to Uncle Sam to cover both money owed above my quarterly withholdings, and payment of the 1st quarter withholding installment. Still, it was a very interesting exercise, and I did learn what to charge in the future, and to carry myself with a touch more of decorum in legal situations.

FORE!

Right after I finished licking my IRS wounds, I received a call from the same Lawyer, apparently willing to give me another shot. "I want to talk to you about a possible patent dispute case involving a golf club design". "I'll be happy to listen so long as you know that if I get on the payroll, my fee will be $300 per hour consulting and $400 while on trial or deposition". He had no trouble with that, but added that for the nonce, he would ask me to work 'pro bono'. "Let's talk", I said, having really enjoyed my just previous outing as an 'expert witness'.

We arranged for a get-together to discuss the details of this new but still tentative case. It involved a truly beautiful club; an iron that came with an equally slick velcro – attached cover. He handed me an information folder with my name printed on it in large letters and then reviewed with me my previous history as a golfer. I confirmed my alleged statement that I had consistently shot in the high 70's when I was a teenager, but gave up the game as a serious avocation when, still as a teenager, I began to throw clubs or wrap them around trees. Satisfied that I could now qualify as both a technical expert witness and a former player of shady notoriety, he told me his problem and how I could help.

The club that he gave me was of conventional design on the hitting edge side of the head. The design innovation was the addition of a trailing T-shaped counterweight with the bottom end of the 'T' attaching to the rear side just in back of the central 'sweet spot' where the ball was meant to be hit. The top bar of the 'T' was aligned parallel to the club's ground-hitting surface. It was different from any club that I had ever seen, albeit I'm 20 years behind the golfing times. I do watch the PGA and the Master's tournaments over TV but, sitting on the old viewing couch, you can't see the design details of the clubs being wielded by the Pros.

The club's designer claimed that because of its then unique design, it would deliver a more repeatable and truer line towards the cup than conventional clubs, even if the ball was stroked outboard or inboard of the 'sweet spot'. He wanted me to opine if I could see any scientific reason why this might be true. If so, he might be able proceed confidently in trying to finally get the patent granted – five long years after it had been applied for. Then, if this happened, he could then bring to court a well known club maker, who had later applied for and received a patent on a design that used similar new design features. This club was a big seller; so mucho dinero could be involved in royalty payments and, more important – I could be paid for my previous pro-bono services, and with it suffer yet another IRS indignity

The Lawyer explained that shortly after filing for the patent, the inventor took ill and did not get back a full head of steam until recently. During his illness, the Lawyer continued prodding the patent application, which had apparently been put at the bottom of the Examiner's work pile. While this was proceeding, the other company applied for and obtained a patent for a similar design in which the 'T' was replaced by a T whose top bar was curved like a horseshoe instead of being straight. And, their patent made no claim that the club would produce true shots even if the ball wasn't hit exactly on the 'sweet spot'.

When the now-recovered inventor finally got his fighting strength back, he asked the Lawyer to again whip into serious action. Although he was most gung-ho about pursuing the case, he was advised that proceeding would be an expensive and possibly non-rewarding adventure. The Lawyer determined that the patent application had become stalled when a patent Examiner made a finding that "there was nothing new" in the original patent application and cited a previously granted patent in which a trailing 'U' piece was used instead of a 'T'. In this earlier patent, the two upper legs of the 'U' were rigidly attached to the back of the club head fore and aft of the sweet spot. The Examiner's supervisor, probably recognizing that there could be a difference in the operating principle between our inventor's design and the cited earlier U – design patent, disallowed the 'nothing new' finding and invited the Lawyer to make a protest and a reapplication, with the original filing date remaining intact.

This is where he was now – asking me if I could first clearly demonstrate the complete dissimilarity in the two patent descriptions and write up my findings stressing why the inventor's idea would really do the job he claimed it would. Later, assuming success in finally getting the patent granted, I would be an expert witness in any subsequent legal action seeking redress; and big money might be had. I told him I'd take it under advisement and would actually use the club myself in a proper setting, and keep a record of my time and expenses in case I lived that long. I kept that promise assiduously!

My highly skilled engineering method to prove the club's claimed operating superiority consisted mainly of bringing the club and its sales brochure along on a couple of my twice-weekly sailboat outings with my crew of old farts, most of whom are retired or active aerospace engineers. One crewman, in particular, is a structural dynamics expert, and he was the one I looked to for the most help. Together, we constructed a scientific-based scenario which is sufficiently plausible to win the first battle of getting the patent. I told the Lawyer that if there were a subsequent court case, the scenario would have to be proved by a very sophisticated computer program analysis which, to have any credence, would have to be done by a specialized company. It would entail an expensive study. Again, in anticipation of future rewards, I naturally kept track of my time on the high seas as well as concomitant boat expenses.

Before I passed on my findings to the Lawyer, I tried out the club on the local Los Verdes Country Club. I tried it, and several friends tried it. The latter all liked it very much, but agreed it was very subjective to know what happened when they deliberately hit away from the sweet spot. They even noted that a different sound emanated from the club than the sound that came from their standard clubs. For myself, not having golfed for at least 10 years, how would I know? I tried a few shots; hitting on the sweet spot first, and then ahead and behind it. I could not detect a discernable difference. Still, there was a structural dynamic physical principle that the club's corrective reaction could be based on. So, backed by the combined strength of my crew's best judgments, I told the Lawyer that I could, in theory, support the claim that a 'truer' shot would be likely when the ball was hit off the sweet spot. He told me to stand by and wait for developments.

I stood by for almost two years, when finally the Lawyer told me the club's inventor had died and his family had decided to abide by his earlier decision not press on. I later heard that a significant part of the story has been left untold: how the Examiner kept coming up with lame excuses for delaying and denying the patent, while at the same time handing and granting the later patent application. It seemed a bit suspicious?

The caper left me with one fine club to add to my estate and pass on to a friend who is a good serious golfer. More interesting, it left me with an idea of how to further improve the club's performance by making some basic design changes: having the shaft attach to the head right over the sweet spot; and allowing the length of the stem of the counterweight be adjustable by its owner, to account for an individual golfer's normal playing habit of missing the sweet spot by roughly the same miss distance. This treatment could be applied to all Irons. For Woods, the design would entail mounting the perhaps 'heavied-up' aft portion of the head on a lockable screw drive to adjust the distance to one's liking. Let the games begin!

ONE POTATO; TWO POTATOS; HOT POTATO

About six months before the untimely death of the golf club inventor, the same Lawyer asked me to make a call to one of his clients to see if I could help him out in a Consulting mode. In doing so, I uncovered a big business that I would never have thought existed - even in my wildest dreams: The construction. Shipping, and maintenance of containers which transport hazardous materials!

I made an appointment to talk it over at their plant. The company I went to consult with lives in quiet obscurity, unbeknownst to the general public except for users and a scrutinizing government Department involved with national security. I was briefed about their problem by their President; obviously a man who knew his business back and forth - and then some. Nevertheless, for mostly political reasons, he felt he required a top-flight designer/analyst to confirm his just-conceived container re-design approach, that cleverly and efficiently solved a newly- formulated Government specification. He also thought he needed a 'Front Man' with the proper credentials to present his new design to government officials, whose main objective in life survival appeared to him to be to protect their asses so that no possible calamity could ever be ascribed to their negligence. He felt that he, himself, was not particularly well-liked by the Agency people he interacted with and that an 'out-of-towner' might be better received. There was no doubt that I had the credentials: a former university engineering Professor, a Professional Engineer, and a former holder of a pertinent high level security clearance. It was clear to me that I could nicely perform the 'Front Man' function; and I told him that I knew just the person to do the heavy analytical lifting which I was not capable of doing. He agreed to take us both on.

The hazardous materials in question were presently being shipped to users around the world in containers made by my sponsor's company and a few other outfits – all under the tight security that goes with the highest level of National Defense. The materials were stashed in a protective core, so the purpose of the shipping containers was to assure no harm ever came to the core. The containers – for size, think of three to four times a typical beer keg but cylindrical, rather than barrel shaped - held the thick walled hollow lead cores in which the materials were shipped. The presently used containers had to meet test requirements that were already mind boggling: The container/shipper had to protect the cores from melting, puncturing, cracking or deforming under the stress of combinations of airplane crashes, deep underwater submersion, raging fires, extremes of heat and cold and rain/humidity, and drops from high places on to spikes and hard surfaces, among others.

There were many authenticated containers presently in operation, and these were re-usable after delivery and generally had a long shelf life. However, a recent pending change in the 'rules' required either a new design, with the possible consequence of either shelving of all the authenticated containers, or making a modification that would pass muster. The forthcoming revised Specification change consisted of survival from a drop onto a spike that had now grown to about 6 inches in height, with a 4 inch diameter. My sponsor had invented a simple 'modification' idea that should work beautifully, but he then believed that he needed an 'outside' expert analyst/designer to confirm his design, as well as run a "Dog and Pony" slide show to convince the Government customer of the righteousness of the approach.

That's what we talked about in our opening discussion, at the end of which I told him that what I thought he really needed was two people; and that I could head up such a team. I had in mind a friend, a former co-worker at TRW and a sailboat crewmate, – a better designer and stress analyst you would be wont to find! Not only that, but he had previously worked with hazardous materials at TRW and, at that time, also had the proper security clearance. Our sponsor bought the package; we agreed on hourly rates based on my newly-discovered knowledge of what the traffic would bear, and agreed on a formal contract which described the scope of our proposed work. I was to prepare the formal presentation and my cohort was to prepare the design modifications and the analyses and test plans that would prove the efficacy of the new design.

Our sponsor's design idea was one that he thought would not only allow his company's containers to be modified at relatively small expense and in good time, but also those of his competitors. What a coup if he could bring this off! To do so, he had to convince what he perceived as a very skeptical Government agency office – who would pay nothing of the cost of the re-design and the requisite rigorous testing program spelled out in the authenticating 'bible'. All this would come out of the Company's pocket; but he dare not proceed without the tacit blessing from the Agency that his approach was viable in their opinion and could be successful. That's why he thought he needed we two 'hot shot' Consultants: to perfect his design idea and conduct a winning 'Dog and Pony' Show in Washington before the highly conservative skeptics.

After a final 'kick off' meeting with our sponsor, in which our proposed tasks and labor division were again agreed upon, and weekly meetings agreed upon, we started on the journey, with a 3 month goal in mind. Since the new Specs would not come into effect until a year hence, we thought we could make the pitch in the 4th month, and with the expected favorable response coming shortly thereafter, our sponsor would have at least a half a year to make one or two samples to undergo testing. Consulting/Supervising at the testing – to be done mostly at a Lab in Texas where they had done previous trials – thus became another addition to our contracted tasks.

Almost immediately, things began to get muddled – probably because of conversations, which we were not privy to, that our sponsor continued to have with his Government customer. The first change was that we were to propose two classes of containers: one for one type of material and the second for another type. This, of course, effectively doubled my cohort's work and slightly increased mine (I had to tap dance faster and funnier). It also raised pay and schedule questions, but our sponsor essentially said that money was not his problem. It was then we realized what a big business this was!

I began creating view graphs meant to lead the customer by the hand into the design process analysis, solution, fabrication and test, which ended with anointment with holy water in one year. My cohort's task was even more creative since it involved hard numbers and difficult mathematical processes. Our sponsor had agreed that, for preliminary design purposes, his 'back of the envelope' calculations – actually very sophisticated – would do, but for the final show we needed computer program verification. After some searching, I found a very appropriate and pictorially beautiful program run by a small, probably one-man, company in northern California, who was anxious to help.

The containers that had been used to shuttle the hazardous materials around the world – under close government supervision – were built up around the roughly one foot diameter core that contained them. The end result was a large 'tin can'- looking steel container. The containers had grappling fittings welded at appropriate handling locations and were painted battleship gray. Inside, shielding the core from spike penetration and inward crush of the steel protective cover, were combinations of plywood, balsa wood, and other materials that charred, but didn't burn, when exposed to fire and/or very high temperatures. These 'inside' cushion materials had the property of being crushable in some directions and reasonably stiff in other directions, according to their grain.

Our sponsor's design innovation 'break-through' was to attach a 'cap' at each end of the barrel whose diameter was larger than the present diameter, which further extended the length of the barrel at each end. Thus an added 'stand-off' distance would be added in all directions to accommodate the new spike size -requirement. The same approach should be applicable to the competition's containers, so long as they met all other requirements.

My cohort not only designed the end caps, but also made improved designs for the two new units that would be required if you started from scratch. While he was doing this, I was busy producing viewgraph after viewgraph, leaving copies of them off after our weekly conferences. The "Show" was becoming an epic – easily one hour's worth without questions. The months went by and deadlines were missed and payrolls became significant. We now worked with our sponsor's Chief of Quality Control – a knowledgeable engineer, who was also a fellow sailor. But the goal became more and more elusive. One day, I was told that my viewgraphs were sufficient and that they would take it from here. After my dismissal, my cohort was consulted from time to time about his designs, but that, too, eventually stopped. I can't tell you today what was/is the outcome. I presume a happy ending, since the idea was basically very sound. But the caper again wreaked havoc with my retiree's income tax bill!

A MEDIA DARLING

Star of Stage, Screen and Radio

While Professoring at Iowa State in the 70s, I appeared on TV or was heard on Radio and was written up in the Ames *Trib* a few times. My pontifications then had more to do with my extra-curricular activities of presenting traditional jazz séances, called "Serenades for Mouldy Figges" than with any academic news. I was a media darling only because the ladies who ran the network TV talk show (the State's NBC TV outlet was located on campus), the radio talk show and a newspaper weekly columnist were all personal friends.

A long dry period of non-notoriety ensued, but after I retired from industry and teaching in 1996, business picked up and I had 3 individual '15 minutes worth of fame' opportunities! The first was a national TV shot on the *History* channel having to do with the propulsion system of UFOs; next came a WEB video on the *SPACE SHOW* having to do with my then new book, "On the Cutting Edge"; and my last – May 2010 – also on WEB video – recording my Technical Society's Distinguished Lecture, "Space Engineering Adventures in the 60s" as presented to the Palos Verdes Seniors Association (www.pvseniors.org). The fun is in the telling:

UFOLOGY

One day in early 2006, Luke of WORKAHOLICS Productions called me to ascertain if I would be interested in participating in a HISTORY Channel production he was doing which had the premise: If there are Flying Saucers or UFOs, how would they be designed? I have always been a UFO fan, and in my early '50s years in New Mexico, where sightings were very prevalent - almost daily – was almost a true believer. Later my enthusiasm waned. I figured if they were so smart, why didn't they announce themselves – or maybe they just didn't like what they saw. In any case, I agreed and we had a good session at our house where they interviewed me in front of a big green backdrop curtain. When the show came on TV, I was sitting in front of a very avant-garde scene – the magic of today's technology.

CATCH A ROCKET PLANE

IN THE FINAL VERSION THAT APPEARED ON TV, THE GREEN BACK GROUND HAD WONDROUSLY BECOME A STAR- AND PLANET- FILLED WORLD WHERE ALIEN LIFE FORMS COULD LIVE IN PEACE, COMFORTABLY

Luke, foreground right, is interrogating me. During the filming process, they threw a number of questions at me. Here is a bit of the repartee, to prepare you for the final production:

Questions posed to an 'Aerospace Engineer'

Q: Are there any aerospace vehicles today that have the flying abilities attributed to UFO's operating in earth's atmosphere?
  A: No, not in one vehicle – etc. etc. (conclusion, UFO's are not terrestrial)

Q: Do aerodynamic factors have any impact on the shape and size of UFO's?
A. Generally NO, but the disc shape - flat on – is very good for relative low reentry temperatures, etc., etc.

Q: What does determine the shape and size of a UFO?
A: Generally, the propulsion system and materials, etc, etc.

Q: How are UFO's propelled and controlled
A: Obviously there must be directed thrust; emanating from jets or interactions with a universal field or fields (eg, gravity, magnetic, etc.

When I was told of the dates of the program, I broadcast a Flyer on the WEB about my forthcoming media appearance via E-mail:

"Dear Family and Friends

   On Feb. 6, at 8 pm on the HISTORY channel, I will appear in a major role in the guise of an 'expert' Aerospace Engineer on a very interesting program; "Alien Engineering". My role is to discuss the shape, aerodynamics, propulsion, and flight characteristics of 'sighted' UFOs.
   *My first TV appearance was in the 70's as hype for my then forthcoming musical opus "Serenades for Mouldy Figges" lectures at Iowa State University. My Feb. 6, HISTORY CHANNEL (8 pm EST/PST, 7pm CST) appearance deals with a subject I know less about (namely UFOlogy) but should be interesting and a hype for my forthcoming (finally!!) first published book, "**On the Cutting Edge**" which should be out late Spring/ Early summer.*

   I'll kill you if you don't watch!

CATCH A ROCKET PLANE

Love Dad or Bob (whichever fits best)"

Before they had appeared at our door to record, Luke sent me a working script; that I worked on and hopefully improved:

WORKAHOLIC PRODUCTIONS (Matthew Hickey and Luke Ellis)
Present

"ALIEN ENGINEERING"

Airing on the History Channel:

Part 1: Premiers Monday February 6th, 2006, 8:00pm E/P and
7:00pm Central

Part 2: Premiers Monday February 13th 8:00pm E/P and 7:00pm Central

Thank you to everyone who helped make this fun show a reality. The two-part program might be described as a cross between a NOVA special and a "Twilight Zone" episode. There's plenty of hardcore science, a bit of UFOlogy and dash of science fiction. We hope you enjoy the show.

Show Premise:

THIS IS AN EXERCISE IN IMAGINATION. PREPARE YOURSELF. SUPPOSE AN ALIEN SPACECRAFT HAS CRASHED IN THE DESERT... WE RECOVERED IT. NOW WE'VE BEEN GIVEN THE TASK OF DECODING THE TECHNOLOGY AND FIGURING OUT

AIN'T RETIREMENT GRAND?

HOW
IT RUNS.

AND, MAYBE IN THE PROCESS, DISCOVER THE SECRETS OF THE UNIVERSE... THROUGH "ALIEN ENGINEERING."

Part 1 includes:
A discussion of UFOs and aerodynamics,
A trip to Lemoore Naval Air Station's human centrifuge to discuss inertia,
A discussion of anti-gravity and electromagnetic force fields,
A trip to the National High Magnetic Field Lab,
A discussion of faster-than-light travel and Einstein's problems with traveling that fast,
A discussion of theoretical wormholes and space warps.

Part 2 includes:
A discussion of antimatter and antimatter propulsion,
A trip to FermiLab where they make antimatter,
A discussion of real-world laser weapons,
A trip to Kirtland Air Force Base,
A discussion of cloaking devices and force field generators,
A trip to see a working plasma generator,
A theoretical discussion of a teletransporter (think Star Trek)
A trip to the SETI Institute, where scientists are searching the heavens for alien signals.

Narrator Voice Over (VO:)
THERE IS NOTHING WRONG WITH YOUR TELEVISION SET. DO NOT ATTEMPT TO ADJUST THE PICTURE. FOR THE NEXT TWO HOURS YOU <u>must</u> SUSPEND YOUR DISBELIEF!

CATCH A ROCKET PLANE

An alien spacecraft has crashed in the desert. We recovered it. Our job: Figure out how it works. Reverse-engineer the technology.

WITH THE HELP OF LEADING PHYSICISTS, ASTRONOMERS, AND ENGINEERS, WE'LL DECIPHER UFO TECHNOLOGY: INERTIA CANCELLERS. ANTIGRAVITY DEVICES AND WORMHOLE EXCAVATORS
AND, MAYBE IN THE PROCESS, DISCOVER THE SECRETS OF THE UNIVERSE.

JOIN US NOW ON UFO FILES: HOW TO BUILD A FLYING SAUCER.

ACT I

VO:
WHAT YOU ARE ABOUT TO PARTICIPATE IN IS AN EXERCISE IN IMAGINATION. BUT *not* AN EXERCISE IN FANTASY.

THIS PROJECT IS BEYOND TOP SECRET. PREPARE YOURSELF.

VO:
This is the alien space ship that crashed in the desert some years ago. We've been given the task of decoding the technology and figuring out how it runs.

VO:
WE'VE ASSEMBLED A TEAM OF SOME OF THE SMARTEST AND MOST IMAGINATIVE RESEARCHERS, PHYSICISTS, ENGINEERS AND ASTRONOMERS AS ADVISERS.

VO:
STEP-BY-STEP OUR EXPERTS WILL TAKE US ON A TECHNICAL TOUR OF

AIN'T RETIREMENT GRAND?

THIS SHIP, FROM ITS CONTROLS, TO ITS AERODYNAMICS, TO ITS PROPULSION. *(should we add biological –i.e. 'human' factors?) (We did)* WE'LL LEARN HOW IT TRAVELS THROUGH SPACE AND HOW IT MANEUVERS IN THE AIR. THIS TASK WON'T BE EASY, BUT IT WILL BE REWARDING. ALIEN TECHNOLOGY IS CENTURIES...MILLENNIA.... AHEAD OF OURS. THIS SHIP IS LIKE A CHEAT SHEET FOR ADVANCED PHYSICS.

VO:
LET'S START WITH THE BASICS; THE SHAPE AND FLIGHT CHARACTERISTICS OF THE SHIP. OUR CRAFT IS A TRADITIONAL SAUCER, BUT EXAMINING RECORDS OF UFO SIGHTINGS OVER THE LAST FIFTY YEARS, IT'S OBVIOUS THAT THERE MAY BE SEVERAL DIFFERENT DESIGNS.

PHYSICS EXPERT

Author "UFOs, What if they're real?"

When you start researching UFO sightings you quickly realize that there are many commonalities throughout the reported cases that stay true throughout all the various reports. The number one thing is when we think of a UFO, we think of a flying saucer. It's a disk shape, usually metallic, glistening in the sunlight -that's probably the most prevalent sighting. Sometimes you have the cigar shaped UFOs, but there are other popular types of sightings as lights in the sky: the glowing orbs that are seen sometimes going very slow, sometimes seen going very fast that are usually reported.

VO:
A TRIANGLE OR BOOMERANG IS ANOTHER COMMONLY REPORTED SHAPE.

CATCH A ROCKET PLANE

VO:
SINCE ALIENS MAY COME FROM ALL OVER THE GALAXY IT MAKES SENSE THAT THEY MUST HAVE DIFFERENT MAKES AND MODELS JUST LIKE OUR AIRCRAFT. EACH HAS ITS OWN PURPOSE AND HANDLING CHARACTERISTICS.

Bob Brodsky
Aerospace Engineer (*first appearance on screen, with Cosmos as back ground*)

When you think of a light plane, the wings generally have blunt leading edges. But the minute you get into higher speeds then the wings have to have sharp leading edges. When you get into hypersonic speeds, you might not have any wings at all. The hypersonic vehicle that NASA is now testing is essentially a flying slab.

VO:
HOW DOES SHAPE AFFECT THE AERODYNAMICS OF OBJECTS FLYING IN OUR ATMOSPHERE?

Bob Brodsky
Aerospace Engineer

When you think about a conventional airplane, generally the forces that act upon it are lift, a force that's provided by some surface, usually a wing. Now acting against lift is the pull of gravity. In the fore and aft direction, the drag caused by the air moving over the surface tends to impede forward motion. That has to be overcome by some kind of propulsion or thrust to counteract the drag.

VO:
FEW OF THE REPORTED SIGHTINGS MENTION WINGS, TAILS, JETS OR PROPELLERS. SO HOW WOULD AN

AIN'T RETIREMENT GRAND?

OBJECT LACKING THESE ESSENTIALS STAY ALOFT AND BE CONTROLLED?.

Bob Brodsky
Aerospace Engineer

You know there's a saying, with enough propulsion, you can fly a brick and that's true.

VO:
BUT ACCORDING TO OUR AERONAUTICAL ENGINEERS, A FEW DESIGNS WOULD BE MORE FLIGHT WORTHY IN OUR ATMOSPHERE THAN OTHERS.

THE BOOMERANG IS A GOOD AERODYNAMIC SHAPE.

WE EVEN HAVE OUR OWN BOOMERANG SHAPED AIRCRAFT SUCH AS THE B-2 STEALTH BOMBER. IT'S KNOWN AS A 'FLYING WING' DESIGN.

Bob Brodsky
Aerospace Engineer

A flying wing is a beautiful shape, at least for subsonic flight, because it's all wing and it's all efficient. The minute you start putting engines hanging out, the minute you put a fuselage, the minute you put a tail, rudder, elevators, you're adding drag. A Flying Wing doesn't have that.

VO:
THE FLYING WING DESIGN DATES BACK TO THE EARLY HALF OF THE TWENTIETH CENTURY. NORTHROP AIRCRAFT COMPANY DEVELOPED A LONG RANGE BOMBER DURING WORLD WAR II.

VO:
COINCIDENTALLY, THE SIGHTING THAT STARTED THE UFO PHENOMENON AT ROUGHLY THE SAME TIME ALSO DESCRIBES BOOMERANG SHAPED OBJECTS.

VO:
ON JUNE 24TH, 1947 KENNETH ARNOLD WAS FLYING IN HIS SINGLE-ENGINE CESSNA IN THE CASCADE MOUNTAIN RANGE OF WASHINGTON STATE, WHEN A FLASH CAUGHT HIS ATTENTION. OUT HIS WINDOW HE CLAIMED TO SEE NINE METALLIC OBJECTS FLYING IN A LOOSE "V" FORMATION. AND MOVING AT A "TERRIFIC SPEED" IN HIS WORDS.

IT WAS A STARTING GUN OF SORTS. INTERVIEWED BY BOTH THE NEWS MEDIA AND THE FBI, ARNOLD USHERED IN THE MODERN ERA OF THE UNIDENTIFIED FLYING OBJECT WHEN HE DESCRIBED HIS OBJECTS *FLYING* LIKE "SAUCERS SKIPPING ACROSS THE WATER." HOWEVER, HE DIDN'T SAY THEY *LOOKED* LIKE SAUCERS.

PHYSICS EXPERT

Arnold later made some sort of a sketch for someone that described the shape and it turned out it wasn't exactly a saucer, it was something more like a badge. But the term has stuck.

VO:
COMPARING ARNOLD'S SKETCH WITH A STEALTH AIRCRAFT FROM THE U.S. AIR FORCE REVEALS THE SIMILARITIES. MAYBE THOSE ALIENS WHO BUILD TRIANGULAR SPACE SHIPS ARE ALSO CONCERNED ABOUT DRAG WHEN FLYING IN OUR EARTH'S ATMOSPHERE.

AIN'T RETIREMENT GRAND?

VO:
NEXT, WE'LL EVALUATE THE AERODYNAMICS OF OUR CRAFT'S SAUCER SHAPE.

PHYSICS EXPERT

That's not a particularly stable shape. It doesn't have any particular control surfaces like the fins or the flaps on a conventional airplane - For changing direction - Either climbing or turning. So, it's really hard to see why this would be a particularly advantageous shape for flying inside any kind of atmosphere.

Bob Brodsky
Aerospace Engineer

People have dabbled making flying saucers. In fact they've shown it can be done. Generally they're saucer shaped, they have a big fan in the center, and they take off and they fly and it's a good show. But it's a lousy airplane. It's a very inefficient lifting surface and they're getting most of their lift from the fan that is blowing air down more like a helicopter. If you were marked, graded for efficiency, it would be at the bottom of the scale as far as a lifting device is concerned. You would do better with a barn door.

VO:

SO WHY WOULD THE ALIENS BUILD SUCH INEFFICIENT AERODYNAMIC CRAFT?

Bob Brodsky
Aerospace Engineer

CATCH A ROCKET PLANE

When you look at what's been reported. Their propulsion systems must be so powerful that it really doesn't matter what shape they are.

VO:
THIS IS THE KEY POINT. OUR ALIEN SHIP ISN'T DESIGNED FOR AERODYNAMIC TRAVEL. THE EFFECTS OF AIR AND GRAVITY AND INERTIA DON'T INFLUENCE IT. SO THE SHAPE OF THE SHIP DOESN'T REALLY MATTER TO OUR ALIENS. AERODYNAMICS ARE OBSOLETE.

VO:
OUR NEXT CHALLENGE IS TO FIGURE OUT HOW THE ALIENS ON BOARD SURVIVE THE FORCES OF ABRUPT CHANGES IN SPEED --- VIOLENT FORCES THAT WOULD RIP APART OR CRUSH A HUMAN BEING.
IN OTHER WORDS, SOMEWHERE ON THAT SHIP, IN THE GADGETRY OR THE CIRCUITRY OR THE ENGINE, THERE'S AN INERTIA CANCELER! WE'LL SEND IN OUR TEAM TO FIND IT.

END ACT I

The show itself turned out to be a good and interesting one, and I thought I looked and acted like a real pontiff. My take on UFOs is now that there indeed were scouting visits, a la Roswell, during the late 40s and in the 50s. I think they came, saw, and just didn't like; and went searching for a better place to colonize.

Luke sent me a TAPE of the whole program. It was done with great taste and ingenuity. He said he would like to use me for future productions. I still jump when the phone rings.

AIN'T RETIREMENT GRAND?

THE SPACE SHOW

In early 2007, I got a call from Dr. David Livingston, a friend who is sole proprietor of *"THE SPACE SHOW"* (www.the space show.com), a WEB-cast wherein he interviews - over the phone - today's featured star of things spacial. He had read my recently published book, *"On the Cutting Edge"*, liked it and wanted to do a program on it. Of course, I agreed and we decided on a date in May. This is how it was announced over the WEB:

THE SPACE SHOW®

*Newsletter and Program Guide for
the Week of May 7, 2007
Now Heard in more than 50 countries!*
Hosted by Dr. David M. Livingston
Web: www.TheSpaceShow.com
E-mail: drspace@thespaceshow.com

- - - - - -

5.   The Sunday May 13 Space Show program from 12-1:30PM PDT welcomes Dr. Robert Brodsky to the show. In addition, Happy Mothers Day to everyone. Robert F. Brodsky is a pioneer in both spacecraft design and the teaching of astronautics. Until his retirement in 1996, he was a professor of aerospace engineering at the University of Southern California, and he has held executive engineering positions at Sandia Corporation, Aerojet, Convair, and TRW Space and Technology. He was named Outstanding Aerospace Educator and University Professor of the Year, among many honors. His new book is "On The Cutting Edge: Tales Of A Cold War Engineer." Listeners can talk to Dr. Brodsky or the host using toll free 1 (866) 687-7223, by sending e-mail during the

program using dmlivings@yahoo.com, drspace@thespaceshow.com, thespaceshow@gmail.com, or by chatting on AOL/ICQ/CompuServeChat using the screen name "spaceshowchat."

About a week before the actual WEBCAST, I talked with David over the phone and we discussed how the hour-and-a-half worth of interview would go. We agreed that the main topic would be the contents of my book. I agreed to send him a Bio and an outline of the topics I would talk about, which comprised:

Topics of Discussion: May 13, 2007

(ALL OR SELECTED)

Chapter 2   Mexico under siege (p. 34)
            Enabling the Space Age (p. 56)

Chapter 4   Space lifeboats (p. 92 0
            Sea launch (p. 105)

Chapter 5   Iceberg utilization P. 130)
            'Inventing' Astronautics (p. 144)

Chapter 6   Cyberspace U. – tie in with previous topic (p. 172)

Chapter 7   The Hydrogen Economy (p.186)

Chapter 8   The Glory of Engineering – CPFF vs FP (p. 196)
            Great Events (p. 206)

On the fateful day, with about 15 minutes lead time, I called the number and talked to David about how the program would proceed. He told me that his show was now 6 years old and that this broadcast would be #713. He said that listeners would be calling in with questions or comments during the show, and that he would relay them to me. There would be two short 'breaks' during which he would be looking over call-ins, while a pre-recorded advisory message about future programs would be heard. He was calm- I was calm - and looking forward to testing my ability to 'think on my feet'.

The actual show went swimmingly. I re-listened to it as I was writing this story and found that my opinions- which were many – have held up to the test of time. We covered most of the topics on the book Chapter list above. You can hear the broadcast by going to www.thespaceshow.com and finding the show by seeking the week of May 7, 2007. It's an entertaining hour and a half - guaranteed!

THE DISTINGUISHED LECTURER

My technical society, the AIAA (American Institute for Aeronautics and Astronautics) sponsors a number of "Distinguished Lectures", which it lists for consideration by its Sections throughout the world. If a Section likes a listed topic and speaker, it requests permission from its near - Washington, DC, headquarters to invite its 'Distinguished Lecturer' to perform at one of the Section's Dinner Meetings. The AIAA covers the travel expenses of the Lecturer, who is a Member, usually of its "Fellow" rank. The list of speakers/topics – about 20 – cover a wide swath in aerospace activities and historical events. Sections normally hold 6-9 dinner meetings during a June-to-June fiscal year and one of those might be a Distinguished Lecture. I belong to the Los Angeles Section (indeed, in the 80s, I was Section Chair) which includes powerful players in the Aerospace scene: Boeing, Raytheon, Northrop Grumman, SpaceX, Microcosm,Inc, and many more, as well as the aerospace academic department student branches at UCLA, USC, and Cal State Long Beach.

After I published my first book, "*On the Cutting Edge*" - 'Tales of a Cold War Engineer at the Dawn of the Nuclear, Guided Missile, Computer and Space Ages', I prepared a lecture for a LA Section Dinner Meeting presentation with material taken from Chapter 4; "*The Fabulous Years at Aerojet - 1958-71*". The decade-plus covered the beginnings of the Space Age and told true 'sea stories' from that era. The Talk was entitled "*SPACE ADVENTURES IN THE 60s*". Somebody heard it, liked it, passed on the word to Headquarters, and, Lo!, I was asked to become a Distinguished Lecturer; an honorific I retain to this day. My next gig, as of this writing, will be at the Redondo Beach Main library in the Fall, for the purpose of 'pushing' this "Rocket Plane" book.

The listing write-up, available to the Program Director's of all Sections, reads as follows:

Abstract: *On the Cutting Edge*, **Space Engineering Adventures in the 1960s**

The advent of exploitation of our new prowess in space flight opened up both the imaginative minds of the pioneers as well as the coffers of a government embarking on cold war one-upmanship with the USSR. No company took better advantage of this environment than Aerojet–General and its newly founded subsidiary, Space–General. The author, first in his capacity of Head of the Technical Staff of the AGC Space Division in Azusa and then as Chief Engineer of SGC in El Monte (1959–1971), got a first-hand immersion in the beginnings of space. He has just written a book, *On the Cutting Edge*, which covers highlights of his career from the late 1940s to the present. The proposed talk will discuss some unusual projects undertaken in the 1960s that are covered in the book. They will include: bidding on the Saturn S-2 Stage; building the OV-3, an early Air Force satellite; the reentry paragliders, IMP, and FIRST; the Surveyor moon walker; an early experiment in sea launch of the Aerobee sounding rocket; and delivery problems of the first DSP (early warning) sensor systems.

The Abstract was followed by a Bio and a picture of the Distinguished Lecturer himself. Since the publication of the Listing, business has picked up - though not at a breakneck speed. I have, in the past few years, given the lecture to the LA Section, the St. Louis Section, the San Francisco Section, the I.O.W.A Section (which I founded, but forget what the acronym means, other than the I.O. stands for Iowa, and – I think – the W. for Western), the Utah Section (Ogden), and to the (local) Palos Verdes Seniors Association. You can catch this act by going to www.pvseniors.org, clicking to Videos, and then, find 'May 2010'

I also developed an alternate Lecture, which I sometimes substituted after conferring with the individual Program Director. It covered stories from book Chapters in addition to Ch. 4. It is called, "*ON THE CUTTING EDGE*" - 'An Engineer's Odyssey through the 20th century, with Stories from the Beginnings of the Computer, Atomic. Missile, and Space Ages'. It's a broader view, but due to time limitations, doesn't dig as deep.

I intend to keep giving the lectures as long as I am physically able. I love to hear the sound of my voice! My latest gig was to the Utah section of the AIAA in Ogden. I did it on a one-day trip. I can't do that anymore!

MY LOVE-HATE RELATIONSHIP WITH 'MACHINES'

Although I have lived in relative peace with them since their inception, I remain suspicious of them and know that they will back-bite me whenever it suits them. Who are 'they'? Why, they are the Devil's seed that we now know as PC's or MACs – computers, by gosh! Actually, I grew up with them. I was a user, in 1948, of the first practical digital computer, the ENIAC – there ensconced in the city of my nativity at the University of Pennsylvania. It helped me in a magnificent fashion to complete my Doctoral thesis – after I had thrown 7 months of effort on an Marchant mechanical computer out the window, having discovered a mistake I had made in the first month of the daily grind.

Thereafter, in industry, I painfully, but of necessity, gave up on the Analog Computers that I so loved, and year-after-year, learned to adapt to this year's new IBM digital monster, which became more powerful and took up less room as each new model supplanted the old. Then, one day in the early 80s, I came to work and found that someone had put an Apple E personal computer on my desk (and my System Engineer's colleagues desks, too) with no further explanation. Together, we figured out how to use them by trial and error. My mistake was - and is – that I never received any formal training on the beast – and I've been paying for that omission ever since!

My lack of patience and my lack of machine finesse would both be highlighted when the evil machines became connected to the outside world by the world wide net. The agony began when I arrived - ON THE THRESHOLD OF THE NET:

Prior to late 1996, when this nightmare began, I had kept my MacIntosh computers in pristine isolation; immune to viruses and web surfing proclivities that come with a hook-up to the outside world. I used my computers as word processors and view-graph composers. When son Jeffrey gave me his MAC SE/30 with HP Deskwriter printer a year prior, I ceded my original version MAC and its Image Writer printer to my wife for her travel journal work. The SE/30 came loaded

with extra features such as Compuserve, many games, and other incomprehensible goodies which a neophyte personal computer jockey could not really be expected to know about, let alone master. But, it did word-process efficiently and faster than the earlier version. I should have left well enough alone!

"You know your problem? You're electronically challenged!" These words from the lips of Michelle, our real estate lady, on hearing for the tenth time why I couldn't send a FAX to her office on a matter dealing with the sale of our Hermosa Beach townhouse. She had made this judgment based on my seeming inability to learn the intricacies of her various paging and beeping devices. My inability to FAX was the figurative final straw. At this point, she seriously doubted that I had ever earned a living as an engineer, and found incredulous a claimed scientific Doctor's degree. I, too, was beginning to wonder. Was I really ready for this new computer-driven world?

Until then, I was happy as a clam in my serene retirement, with my now burgeoning first book taking on a life of its own. In the back of my mind, however, I continues to hear the seductive siren voices of our cousins in distant Haifa and Florida, whom we dearly loved, urging us to go on-line: at least to achieve e-mail capability. They knew my aversion, or fear, of networking which I felt would take up too much time away from my book writing. And, this at a time when its completion was both becoming a race with the grim reaper, as I was already in my 70s, as well as threatening to become a trilogy because of the backlog of stories I wanted to write.

Still, they persisted and I must admit I was impressed by the prospect of being able to receive messages of import from far places in real time. Such messages were now being relayed to me by phone via the ministrations of my then Teaching Assistant, Bogdan Marcu (now, in 2010, Dr. M).

"Why not me?", I thought, especially since Dan operated from his home phone at only the cost of a local call to get on the world-wide system. But, I didn't take the step. I just wasn't psychologically ready yet.

In late November, the usual Wednesday afternoon "Old Farts" crew under my gallant command was looking for the first whale sighting of the Winter sailing season. As it often did, when not discussing prostate, vascular, ague, and potency problems, the conversation turned to the equally esoteric subject of computers. It was then that Mike Kerrigan, the only 'young' – at 57 then – member of our crew uttered the fatal word that initiated this mind-boggling drama. He advised me of an advertisement in this morning's *LA TIMES*, that Fry's, a local large electronics store, was running. It featured a $40 MAC-compatible modem that promised the possibility of sending and receiving FAXes and, he explained, also opened the door to e-mail and web surfing capabilities.

I was enthralled! After consulting with my Mac repair technician, who assured me of the apparent compatibility of the new modem with my Mac, I decided to go for it. Disdaining the technician's willingness, for $30, to install the called - for upgrading to Operating "System 7" and the new modem software, I decided to install both programs by myself. My crewman, Myron, agreed to loan me the System 7 installation discs, and the three start-up discs that came with the modem promised user-friendly action. I had no trouble installing the new System 7 Operating System that replaced the less capable System 6.8 that came with the machine. I then carefully followed the step-by-step procedure called for by the three new modem discs, and was rewarded by a "success" message when, as instructed, I registered my modem with the manufacturer. However, when I tried to send my first FAX, only the phone connection was made - nothing else! I spent the next day trying to figure out what happened. Naturally, when I called for technical help, a long distance call to Oregon, I was repeatedly told to expect a 20 minute wait. Even when I called at their 5 a.m. opening time the next morning, I got the same dismal message with little hope of ever talking to a human being.

In mild desperation, I decided to repeat the entire new modem installation process, which happily did lead to my triumphant new ability to send and receive FAXes. Unhappily, I could do nothing else, for when I tried to word process with my Microsoft Word routine, I got an "overcrowded" message and could not continue. By now a state of complete frustration had been achieved, and my bad tempered evil twin had taken over. Thus began the first of the highly voluble anti-machines tirades which would now and forever accompany me through life. The outburst scared the hell out of my wife. She told me that the neighbors would think that I was abusing her.

The frustration continued. I tried dumping the new software hoping to go back to where I had been; foregoing the $40 investment. But, no such luck. I couldn't get back there. At the painful cost of unwiring the set-up and loading it into the car, I delivered it to the repair technician. After two days of waiting, and paying him $10 more than the $30 he had earlier quoted as the price for the installation task, I regained both the ability to 'word process' and to send and receive FAXes. At last, a new start!

But, I had lost a week of time in my new book writing career and clearly suffered a rise in blood pressure - both of which I could ill afford, as I was rapidly approaching the age of 72 and the yet unbeknownst appointment in Samarra that awaited me in the Spring. As I wrote this, I had no doubt that the aggravation and indignity that I had suffered at the hands of this supposedly 'user friendly' machine monster contributed heavily to my soon-to-be discovered heart problems.

Now, seeing a brighter day ahead and encouraged by my instructor in the "Writer's Workshop" class I was taking at Redondo High School, I submitted my first commercial writing effort for publication. It was a piece entitled "On the Threshold of the Net ", and described the tribulations noted above in "750 words or less ", and mused about the future. The smasheroo finish went like this: "Now I face a new far-reaching decision; should I go 'On Line'? In making this decision fear- aye, even stark terror- is foremost on my mind. I can see the possibilities for massive screw-ups, followed by equally massive heart attacks. I further fear that I would become engulfed in net surfing, thus losing more valuable time. On the positive side would be the e-mail capability to friends and loved ones. I shall pause at this threshold, and see what the future brings." Little did I know how prophetic this pronouncement would be!

Our 1996 end of the year Florida trip included visits to my cousins in Miami and Naples, who regularly used email to correspond. While there, we conducted an e-mail exchange with our mutual cousin in Haifa. This convinced me to plod on, damn the torpedoes! On return, the next step was either to take the conventional path and sign up with an outfit like America On-Line, which already had sent me an installation disc, or to tie into the USC system, which would only cost a local phone call every try. Naturally, I opted for the latter and asked my sole remaining post-retirement graduate student to find and refer me to an expert colleague to help get me established.

Along came Music Department senior and computer nerd, Steve Wagner. Why Steve took on this job, I'll never know. Surely it wasn't for the money, since he was happy to accept whatever I offered for his time. He seemed ideal for the job since he had exactly the same equipment I had, along with the requisite SoftWare to connect me to the USC system. Steve, however, never had the time to consecutively put together a successful work session. Just when I thought we had accomplished something, he would have to leave, while offering the advice to "try it out". It never worked! I was not yet 'on line'. My fuse continued to get shorter.

A couple of weeks before I was going to undergo an angiogram/angioplasty procedure, the FAX receiving ability, for no good reason, went west. In desperation, with my heart rapidly failing despite the fact that I had no symptoms, I called for "The Computer Geek", who also happened to be a 'Steve'. We then had a $60 session, during which Steve reloaded all the software, trashing all the folders that had to be trashed.

This session, still not successful, included him talking to the modem people in Oregon. They concluded that the modem was not right, and said they would send me a replacement and new installation software. After my heart repair and before our move into a newly purchased condo, the original (musical) Steve installed the newly arrived modem. FAX problems remained and the on-line hook-up was also imperfect. Steve then decided that we had computer problems.

So, just before the house move to Redondo, I again packed up the whole kit and kaboodle and took it to the MAC technician. I told him to do what he had to do to get the damn thing working and to take his time, since we were moving. A month and $210 (a drive replacement, apparently) later, I picked up the equipment, hooked it up in our new digs, and after a few false starts, immediately determined that I couldn't receive FAXes, and didn't know enough to also check out the WEB connection by myself. Fearing a heart relapse, I decided to cool it until after my now scheduled mid-June quadruple laminectomy back operation.

A few weeks after the operation, once again able to walk without leg numbness, I picked up the thread by calling

Oregon. After an agonizing and expensive day, we determined that even though everything seemed to be ok, my telephone was not recognizing an incoming call. The technician asked, "You don't have another phone line, do you? It hit me like a ton of bricks! I suddenly remembered that the previous tenant had told me that he had a separate line for his computer equipment! I used a spare double phone socket, and sure enough, I could not only receive FAXes, but I had my own private machine line that I, at least, was not paying for!

Finally, a happy ending, you say. Of course not! When I tried to print out the received FAX, the machine bombed! After another try, it even bombed when I asked to simply view the received FAX. I tried to find Music Steve at his summer retreat, but he said I would have to wait for school to start in late August. My graduate student, for the nth time echoing the sentiments of my other friends who were witnessing this ongoing drama with great amusement, told me that for the money I had already invested in the quest (by now, around $500), I would be well on the way towards the purchase of a modern machine. He wasn't inclined to again find me some help, so I once more called on Steve Geek.

In the $90 session that followed, he decided that there must be some interferences and decided to trash all folders except those essential to the job at hand and reloaded all programs. He left feeling all was OK. But, of course, it wasn't. After a few outbursts of temper and the brief thought of shooting the damned machine, I got back on the phone to Oregon. The first call, lengthy though it was, did no good.

The next call to Oregon connected me to a sadistic lady technician. In merely ten minutes, by getting me to move things into storage, desk top, and/or trash; she wreaked complete havoc with my system! I couldn't even word process! I desperately tried putting the files from storage and trash back into the system, but to no avail. She had signed off by advising me to call Apple and gave me an 800 number to call. She suggested that I tell them that I had an incompatibility problem. When I called the number, it turned out to be the Psychic network. They told me that they could probably help, but I wasn't having any!

Back came Steve Geek for what I vowed would be the last time. I was so cowed by this time that I told him I would settle for regaining my position of 8 months ago with only a word processing capability. Steve soon achieved that, and also restored a working FAX ability, but did not load the use e-mail software. He advised resting on our laurels for a couple of weeks before trying for e-mail capability. Tight as a string, I acceded, with relief.

Two weeks later, having survived two sets of family guests, I called on Bogdan to finally lead me into the promised land, knowing full well that the outcome would more than likely be tragic. After one false start cancellation, he showed up on Saturday, October 11, 1997 - a day that shall forevermore be celebrated in the annals of cyberspace. He loaded the disk that would connect me with the USC e-mail system, using his wife's password and account, but authenticating my <bbrodsky@spock.usc.edu> address. When he then tried to go on-line, he was immediately rewarded with an 'insufficient memory' advisory. We determined that there were actually over 8 megabytes of memory available somewhere, if we could decommission some other lurking software.

We called Steve Geek, who came to the rescue by advising us to make sure the FAX software was out of the loop. Sure enough, when that was done all seemed to be OK. We sent and received an e-mail to a friend! On October 12, I triumphantly sent e-mail messages to our cousins in Haifa and Florida, announcing that I was 'On Line'!

So, almost a year and $600 later, with an associative heart blockage to boot, I finally had the ability to send and receive FAXes and e-mail, and still could word process as before. A total victory over the non-user-friendly enemy! The day was won and my Mac was saved from obvious premature antiquehood.

When I read this story to my then newly-joined 'South Bay Writer's Workshop' group, I advised them that machines were the work of the devil; and related to them my tales of woe suffered in dealing with inanimate objects. They sympathized; one of them related machine truculence with the psyche of his ex-wife.

My troubles continue to this day, even with the switch to a modern PC. The machines have a mind of their own and do terrible unexpected, unwanted, mean things. Is this the 'Never-Ending Story'?

'ROUND THE WORLD – SPRING '99

It all started in the late summer of '98, when our Haifa cousin Susan and her husband Dick told us that their beautiful talented daughter in Jerusalem was engaged to be wed some time next May, probably at her sister's home in Be'ersheva. Without hesitation, we said we'd be there! Gradually, we began thinking about the trip. Now, my dear wife, Patti, her curiosity whetted by having made a serious illustrated-by-hand-drawing study in elementary school, has always wanted to see the Taj Mahal. So, we started thinking in terms of going around the World, stopping to see the Taj either on the way to or back from the 'Old Country'. In Israel, we would naturally make our headquarters in our old Haifa stomping grounds, and catch up on the many friends that we have made there during out three previous long-term stays; two of them while I was a visiting professor at the Technion.

As the wedding date started sliding towards late May, it became apparent that for hot weather/monsoon-season reasons, we should make the circumnavigation by traveling Westwards. For Taj viewing in the full moon opportunities, a late April departure (actually April 29) was called for, with a one month stay in Israel planned. Naturally, on our way back, we would visit our friends in Germany and France, and our son and his family in Swampscott, Massachusetts.

Our plans started to gel when the wedding date settled down to May 30 and Susan and Dick found an inexpensive apartment for us in the nice Mediterranean - overlooking Carmeliya neighborhood, not far from our previous mountain-top dwellings near the central Carmel section of Haifa. They said that the apartment could probably hold another couple under crowded conditions, should we have any guests. Under the duress of signing a lease now, we established May 6-June 6 for our Israel stay and built the rest of the trip around those dates.

Through the Travel Section of the LA Times, I got in touch with a San Francisco travel agency which specializes in round the world bookings, and with the help of an ace travel agent managed to establish an itinerary of Singapore (via Taipei), New Delhi, Amman, Frankfurt, Paris (via high speed TGV -train of Tres Grande Vitesse- from Aachen to Paris; 350 miles in 3 hours !), Boston, and home for under $2000 per each! By using Royal Jordanian Airlines from Delhi to Frankfurt, we ostensibly saved $400 each over El Al prices- but as you will see there is more to this than meets the eye.

I went on the internet and found that there are two buses a day that do a Haifa / Amman and vice-versa traverse, but I could not find out where you got the bus in Amman. By e-mail, I asked Dick to find out more about the bus, giving him the name "Trust Transportation Co." He wrote back the first of "There is no bus from Haifa to Amman" proclamations, but allowed that he did find a daily Tel Aviv-Amman bus, which obviously solved no problems. We then agreed that on arrival in Amman, I would call him at our mutual friend's house in Afula- near the Jordan border crossing at the King Hussein Bridge (over the Jordan river), and he and Susan would leave for the crossing - the same one we had used two years earlier on our jaunt to Petra - in time to meet us on the Israeli side of the border crossing.

Plans were made via e-mail for our stays and tours in Singapore and Delhi/Agra. We had invited our friends, Jack and Sue Cleland of Sedona, both devout Catholics, to visit us in Israel and get the fabulous Brodsky tour, which would now include some previously neglected Christian holy sites. Then, we carefully planned the packing logistics; schlepping only two large bags with wheels; a smaller strap-on-to-larger-bag rolling bag containing toiletries and valuable papers which we would keep with us on the plane; my camcorder case; and Patti's purse, and a shoulder catch-all bag. The latter always carries special "water" bottles whose Vodka contents we used to "thin" the airlines' iced orange juice. Yea, verily, the veteran travelers were now ready to go on yet another; "This is our last big trip, we're just too old for all this nonsense!"

The trip to Singapore (which is just North of the Equator, and was very hot, but dry) on China Airlines, aided by our "water" supply and sleeping pills, was uneventful. We arrived there in early afternoon feeling pretty good despite having lost a day to the International Date Line (Would we ever get it back? At my age, everyday is important!). We arrived at our hotel, the International YMCA in the heart of the boutique area (an absolutely first class hotel with reasonable prices and a gorgeous view of the harbor), and phoned the regional head of Hughes Communications at the suggestion of his California boss - a friend of ours - and were advised about the best use of our limited time there and the restaurants to try. The advice – which we followed assiduously - turned out to be just right !

It developed that English-speaking Singapore is an upscale Hong Kong, populated chiefly by people of the Chinese persuasion, and is the biggest port in the world (as evidenced by the multitude of large ships standing by in the harbor area waiting for unloading accommodations). Our plan for the rest of the day was to visit Orchard Avenue - the local Fifth Avenue - and to see the famous Raffles Hotel, have dinner, and call it a day. Next day – Saturday - we would visit and explore Sentosa Island; and lastly take a city tour on Sunday before departing, shortly after midnight, for New Delhi.

Raffles, an easy walk from our hotel, had recently been refurbished back to its original British-Empire state of grandeur, and was indeed a grand old lady well reflective of the empire at its peak. We had the mandatory "Singapore Sling" at the Long Bar and found it to be an insipid and cloying drink, hardly an antidote to the hot- around 90 degrees - but dry and therefore tolerable, climate. There is a mall-like area near our hotel called Chijmes, a remade monastery now comprising shops and many excellent ethnic restaurants, where we had our evening meal that night, and the two subsequent evenings.

The next day, we taxied to Mount Faber, and by a very high gondola ride over the harbor, arrived at Sentosa, Singapore's playground island. We took a slow open train which circled the island and noted the stops we wanted to make. Out of many options, we choose to visit an excellent nautical museum and an unusual aquarium (where a moving beltway took you through a glass-enclosed tunnel surrounded by great varieties of sea life whizzing over your head) and returned to the mainland by ferry (much like the famous ferry connecting Hong Kong to Kowloon). We had late lunch at a very authentic restaurant in Chinatown, and did Orchard Road some more on our way back to the "Y". Before dinner, we had cocktails in the observation room of the highest (71 stories) hotel building in Asia - a Westin Hotel where you get a great view of the whole of Singapore.

The tour on Sunday covered some of the ground we had already seen - such as Chinatown and Mt. Faber- but emphasized what a lush, clean city/state Singapore is. We saw the old colonial part - replete with cricket club - the Parliament House, Supreme Court, and City Hall; drove through the Arabic and Little India sections; stopped at a resplendent Hindu temple (Sri Mariamman, the city's oldest, right next to Shakey's Pizza) which featured a tower with hundreds of beautiful colored carvings of figures and cattle, and ended up at the spectacular Orchid Gardens. We didn't realize that there were hundreds of varieties of the flowers - most being lovely. It was a fitting end to our visit.

Shortly after midnight, we embarked to New Delhi on Air India. We encountered a group of people who, according to an official, were traveling for the first time. Their "pushiness" exceeded the best native Israelis could do by far! But, we survived and arrived in New Delhi around 4 am, after a time zone change which was one and a half hours! (There are two such off-hour time zone changes in that mysterious part of the world, and it drives you crazy to find your watch inexplicably disagreeing with local time by a half hour). As scheduled, our driver-for-the-duration, an agreeable young man named Pankage was there, replete with a sign to assure our rendezvous. He subsequently proved himself to be driver with nerves of steel; in a place where such nerves were continually called upon. He understood English perfectly but seemed loathe to speak it in any but basic forms. We immediately felt the high and humid heat. It never fell below 100 degrees while we were there, and reached 115 in mid afternoon. We never saw a blue sky; the sky was white with the Indian summer brand of smog which only the monsoons wipe away, apparently.

AIN'T RETIREMENT GRAND?

We were delivered to our Delhi headquarters, the India International Centre, located in the midst of diplomatic row. Our stay there was arranged by the father of a Friend - a former student of mine - Madhu Thangevelu. His father, Dr. Thangevelu – just deceased in 2009, was employed by the World Health Organization and was a member of the Centre, thus – through his intervention - making us eligible to stay in this lovely hostel-like convention center where people meet to discuss solutions to India's many problems. The compound included about 40 two-room suites, many meeting rooms, a dining room and a bar; all in a sylvan setting adjacent to large park. Madhu says he remembers staying there with his family (his Father lived in Bombay, now Mumbai) and romping around the fields with his brothers and sisters. It also houses a travel agency which arranged for all our excursions in India. We moved in, had breakfast, watched CNN, napped a bit and set out on our 11 a.m. city tour; Pankage and our guide, Jimmy, having arrived precisely on time to pick us up.

Unlike Singapore, a city replete with high rise buildings and an approximately equal 4 million population, New Delhi's highest building might be a 6-8 story apartment house. Except for Old Delhi, and the incongruous cows which wander unfettered through the streets, it is a typical sprawling metropolis with a few outstanding sites. We saw the "India Gate", a huge Arc de Triumph-like WW I war memorial, the Parliament houses and adjacent President's palace, arranged as a U, with an impressive mall leading to the India Gate about a quarter of a mile away. We had lunch with our guide, an erudite educated man who opined that India's vast problems would never be solved by the succession of crooked politicians who were and had been running the country. We proceeded to Lakshmi Narayan Hindu temple, where we were blessed by a Monk who bedecked us with flowers and put a red spot on our foreheads (As you will see, the blessing didn't take); went to the famous "Red" fort where Shah Jehan, the builder of the Taj, moved the government after a severe water shortage in Fatepur Sikri; drove through Old Delhi and called it a day; retiring early to be ready for our early morning pick-up for Agra and the Taj Mahal, and then nearby Fatepur Sikri.

We marveled at Pankage's driving skill as he negotiated the hazards of the suburban towns of Delhi, fending his way through heavy auto traffic in a very nice affordable, always white, sedan which is made in India and is ubiquitous. You seldom see foreign autos, except in the diplomatic/governmental areas – but there are a plethora of bicycles, bicycle taxis, tractors, camels, donkeys, an occasional laden elephant, hordes of brightly clad pedestrians, and the ever present cattle. We also got our first looks of the abject poverty that abounds in India - but nothing like we would see later. As we drove through the city, we saw plethora of seemingly emaciated people, living in huts and filth. Eventually, we got to a semi-freeway, and uneventfully arrived in Agra, a bustling medium-sized city, about 110 miles away. About halfway there, we had stopped for lunch, being careful, as always in India, to drink beer or bottled water. Madhu had warned us of the dreaded "Delhi Belly", whose revenge is considered much worse than Montezuma's, even if Pepto Bismal is administered liberally. For trip preparation, we had taken Malaria pills, and were given booster Tetanus shots, but not Hepatitis 'A' shots, as our doctor thought they would not be necessary in such a short trip, so long as we didn't bathe in the Ganges, or the like.

We stayed at the new Trident Hotel- an oasis insulated from the grim realities of Mother India - located near, but not in sight of the Taj. Almost immediately, our new local guide called us, rarin' to go. We cooled him by saying we wanted to have a leisurely lunch before starting on our tour of Agra. This apparently ticked him off and he remained somewhat surly throughout - the complete antithesis of Jimmy.

We were picked up at 2 pm, and went to Agra Fort, in which Shah Jehan had a palace where his entourage occasionally stayed while the Taj was being erected, and where he spent the last of his life under house arrest - deposed by his oldest son- but at least in view of the Taj, where his dear wife, Mumtaz Mahal, (and later, he) was entombed. The temperature in the fort was a mere 115 degrees and the visit was tortuous. We did get our first look at the Taj, a distance away and obscured by the smog. The glimpse only whetted our appetite. Much to our guide's chagrin, we then asked to return to the hotel, to be picked up later to see the Taj, for the first time, at sunset.

What can you say about the Taj Mahal, after you say it's magnificent! As the sun grayed the normal completely white Indian sky, the white marble took on subtle shades of blue, and the natural beauty of the structure was further enhanced. We had arrived there from the parking lot a quarter of a mile away. At the entry gate, a high walled structure that surrounded the Taj on three sides - the backside faced onto the now dry river bed - via a crazy contraption powered by a motorcycle engine. It carried four comfortably, but we saw some with two extra hanging on to the side ! Our surly guide bought us entry tickets, and collected the rupees from me (not cricket according to our agreement with the travel company) and then shortchanged us! I chose not to call him on it.

*SITTING IN FRONT OF THE
REFLECTION POOL AT THE TAJ*

As we entered, we almost gasped at the shear beauty of the site- the gorgeous mosque with the long reflecting pool leading up to it made the whole trip an instant success! Only pictures can do justice to the site if you cannot see it in person - and we have plenty of them! I took camcorder pictures from the entry way, but had to check the camera before going further (<u>still</u> cameras were not banned). We were almost immediately accosted by a professional cameraman who offered to take pictures of us at a dollar each. We said we would try a few, and he and his aide immediately took charge of posing us innumerable times. We ended up buying a booklet at $40 - the best investment we've ever made as it turned out. When we finally got a chance to see the Taj unescorted, the sun was starting to go down and some of the magic we expected was beginning to happen. Patti used her throw-a-way camera to record.

We soon were approached by a gentleman who said he was only a humble gardener here, but could show us the best views to take sunset pictures. He rushed us from one spot to another, pointing and saying, "here". At the end, I thanked him profusely and offered him what I thought was a nice tip of $2.00. He nastily scorned this, and I ended up paying $5. Scammed again – humble gardener indeed! As we left, we found that due to a terrorist threat a few months ago, the Taj was no longer open on full moon nights - and there is no good way to see it in moonlight due to the surrounding wall. We did buy a beautiful moonlight shot postcard-some compensation? Thus fell a major reason for planning our trip the way we did.

As we went back to the hotel, we discussed tomorrow's activities. We wanted to see the Taj at sunrise, return to the hotel for breakfast and checkout, and the go to Fatepur Sikri before returning to Delhi. The guide started to persuade us that his services were not really necessary in Fatepur Sikri- we could hire a local guide there that he would pay for - and we cheerfully agreed. The Taj at sunrise presents a different, but equally sensational aspect. The color shades and shadows give another breathtaking view. We drank it in and proceeded to Fatepur Sikri, about 12 miles away, East towards Jaipur (which people tell us is also a worthwhile stop). Pankage found us a guide, a weathered old man - older than me! - cloaked in robes and sporting a gold tooth. He was very knowledgeable and had much enthusiasm for the very impressive fort that once was the seat of the Indian government under Shah Jehan. We entered through the 175 foot high gateway that guards the Fort. The insides were very impressive - more so than the Agra Fort and the Red Fort in Delhi. The royal quarters were spacious and – for the fifteen hundreds when they were abandoned- looked to be comfortable. It was also very hot - we had to tell the guide to stop and rest quite a few times (he remained gung-ho!)

and was disappointed when, exhausted, we declined to see the whole shebang.

Since we did not have to take the surly guide back to Agra, Pankage elected to return to Delhi via a "shortcut" through the back roads with which he was not familiar. As uncomfortable as the trip was (it was very hot despite the car's air conditioning), in addition, we lost our way several times; and later badly needed a pit stop (which never materialized), and were hungry. We were glad we had endured the three hour ordeal - for we saw the "real" India. The "real" India is not a sight you really want to see. The poverty, filth, and chaos we saw (e.g., children sharing water holes with herds of pigs, while clothes "washing" was going on) along the way unbelievable. Many people were homeless and many appeared without hope. We saw no evidence of schools or outhouses. The little villages were jammed with humanity and frail living structures. Even the fields looked starved. We gradually lost our taste to ever go to India again.

After wandering '40 years in the desert', we finally got back on the freeway and stopped to eat at the same place as on our way down to Agra. As soon as we arrived at the Centre, Patti made us drinks, using ice cubes from our suite's refrigerator. As it turned out, this was a very bad gaffe. We had dinner and went to bed early, since Pankage would pick us up at 3:00 am to go to the airport. The dreaded "Delhi Belly" hit us both at about 1:00 am, but only - so far – lightly. Its full vengeance was not to be wreaked until just before breakfast was being served on the Royal Jordanian Airlines on the way to Amman. My stomach felt queasy, and I asked for tea. Despite this, I was turning white as a sheet. Suddenly, I knew I had to get to the bathroom, but before I could raise the tray, I vomited all over the place! The chief flight attendant hurried over - she said she knew I was in distress earlier - and cleaned up what she could. I think I set back the mid-East "Peace Process" (in Israel, they pronounce it "Piss", and I have never been sure whether it is an accent or a social commentary) a few years. I cleaned up as best I could, but did feel much better. But I had to sit another 4 hours in soiled, smelly clothes, which I'm sure pleased our neighbors no end.

We arrived in Amman at 9:30 am local time, readjusted our wrist watches for the crazy half hour bit (little did we know that Jordan was on standard time, while Israel was on daylight time) and, with some difficulty but armed with Dinars that I had purchased in India), made the call to Afula to inform our cousins of our arrival and intent to get a taxi to the King Hussein Bridge crossing. We carefully told the taxi driver where we wanted to go and felt sure that he understood. On the way out of the airport, he showed us the only airport hotel, the Alia, and we discussed staying there the night before our early Sunday morning departure to Frankfurt at the end of our stay in Haifa.

The drive to the crossing seemed longer that we remembered from two years ago, but the border station and procedures seemed the same. When, after an hour of customs hi-jinks on both sides, we arrived safely on the Israel side, we wondered where cousins Susan and Dick were. I took the opportunity to change into clean clothes. We waited and waited and were just about to call Afula when we were tapped on the shoulder and asked if we were "The Brodsky's" ? It turned out that the wicked taxi driver had taken us to the Allenby Bridge crossing, the one people going to Jerusalem use. It was some 60 miles south of the King Hussein crossing where our cousins awaited. A $100 taxi ride and two hours later, we were in Afula and reunited with our family and friends, but the savings over EL AL travel were already being seriously eaten into. Our one month stay in the Old Country now began, but the ravages of the Delhi belly would continue for the next few days.

On our way to our Haifa apartment, Susan and Dick told us that they were having some stress. A few days before our arrival, their daughter had called them to say that the wedding was off! Then, next day, it was on again - but they were still biting their nails. In one week from then, next Thursday, the first big wedding event was scheduled. It was to be a party in Haifa giving their friends an opportunity to meet the prospective bridegroom, and to bring wedding presents - since the wedding was to be in Be'ersheva. This was to be followed by a Saturday night fete for their congregation at their nearby Shul. Their uneasiness was enhanced by the forthcoming arrival of two of their sons with their wives, a granddaughter, and Dick's sister; all coming from the States. They had also made extensive plans in Be'ersheva for a rehearsal dinner, an all day guided bus tour to Masada, followed by the wedding and banquet. In addition, they helped their daughter with her wedding gown and other generous pre-nuptial gifts. All this in the face of Dick's imminent retirement from teaching at the TECHNION.

As soon as we entered our new rental apartment, we could see that although it would be great for us for a month, there was no way that Susie and Jack could fit in. There was a living room with a fold-out couch, a bedroom of good size, and the craziest combination kitchen-bathroom that you ever saw! In one corner of the room, there was a shower emanating from the ceiling, and a nearby toilet seat – all surrounded by a curtain on pulleys running through a rail attached to the ceiling– hospital style! In two days, our friendship would have been a shambles.

So, we asked cousin Susan to look for a rental place for them while we checked out local hotels. All of the latter were over $120/night, but Susan found a beautiful apartment for half the price, whose availability fit the Cleland's travel plans. We went to see it, and it was super - much nicer than our apartment, with a panoramic harbor view all the way to the Lebanese border. We grabbed it and proceeded to reacquaint ourselves with our Haifa friends and the beach. That Monday, revisiting the TECHNION, I gave my invited lecture on 'Resolution from Space' to a wildly appreciative crowd that shouted "huzzah" (or, Oy Veh! I forget which) and carried me off on their shoulders.

Before our Arizona friend's arrival, I decided to plan how we would get back to the Amman airport for our Sunday, June 6 flight to Frankfurt. It seemed obvious that there was no way to leave Haifa very early on Sunday with any hope of making a 9 am plane, especially if this was an hour earlier Israel time and the border crossing did not open until 7 am.. It was also worrisome about traveling to Jordan on Saturday, since Israel pretty much closes down public transportation on the Sabbath. I went to a travel agency to buy a ticket for the bus between Haifa and Amman and to make a reservation for Saturday night at the Alia Hotel at the Amman airport, which is a good 25 miles from greater downtown Amman. After considerable rummaging around on the computer, I was told in no uncertain terms that, "There is no bus to Amman" (confirming Dick's earlier pronouncement) and "There is no Alia Hotel"! It became immediately apparent that the tenuous "piss" agreement between Jordan and Israel had not really taken hold. Back at home, I called a number in Nazareth which I had taken from the internet, and was assured that there were two buses daily from Haifa to Amman, and was told where to get a ticket in the harbor area of Haifa.

Susie and Jack arrived on Wednesday, the day before the big pre-wedding party. They had missed Lufthansa connections in Chicago and arrived at Ben Gurion over 5 hours late and without luggage. Before we took them to their apartment, we showed them ours - so that they would see why staying with us was impossible. We had a guilty conscience, since they did not anticipate the extra expense, but they loved their place and their landlord Zvi, a real Israeli old timer, continuously regaled them with sea stories about the beginnings of the country. Alas, it would be exactly one (!) unbelievable week before their bags showed up! The airline allowed them to spend $100 each on clothes- and they managed by doing a lot of laundry and clothes borrowing. Both deserve "Hero of the Soviet Union" awards!

The next evening, after having given up for today on what would be a daily luggage vigil, we arrived in the lobby of our cousin's apartment for the big introductory party, wedding present in hand. Right then and there, we should have suspected the worse. I pushed the bell and nothing happened! We had to wait until a resident came along to open the door before we could enter.

We finally met the groom, a fiftyish sociology professor and musician (clarinet and soprano sax) face-to-face. I had been communicating with him by e-mail because of our mutual interest in traditional jazz and his consummate interest in some of my rare Sidney Bechet records. In person, he was likeable and voluble. He confirmed that he had bought a house in Roq Brun, near Beziers in southeast France where they would live for most of the year, and that his two daughters - from his previous marriage - and his mother were all coming from Old Blighty for the wedding. He also played me a few choruses on his clarinet - but not enough for me to truly judge his prowess. He said he had been living with his bride-to-be in her Jerusalem apartment for about a month - the longest continuous exposure to each other in their one year courtship. We next joyfully talked at length with her, and heard her express her trepidation at leaving her family and friends in Israel for a new country - although she does speak French avec some fluency. She said she hoped that they would live some part of the year in Jerusalem and, to this end, had not put her apartment on the market. She also noted that her man was not enthralled with her flute playing; and hoped he would relent in his insistence that her dog, Jake, now 13, be an "outside-only" pet in Roq Brun.

That night, staying at our cousin's place, the bride and groom had a serious argument that was soothed by Susan and Dick; but after the successful reception at the Shul on Saturday, Susan found her daughter curled up on their living room sofa at 2 am. Her first words were, "I'm not going through with it!", recounting the many reasons why. We were never sure what the incompatibility was – since she had never been married up to this mature state of her life, fear may have been a real factor. At some point, the groom emerged from the bedroom, and was told the news. She told him she was not returning to Jerusalem with him. He tried to dissuade her, but was politely asked to leave. They fed him breakfast and he took off for Jerusalem and returned to England. Could this marriage be saved? Nobody on our side thought so, and even she seemed somewhat relieved.

Now began the many calls to the party attendees to give them the news and to the USA to sons, sister, and granddaughter to inform them that the wedding was cancelled. All replied that they would come anyway to show family support. The Be'ersheva activities would go on as planned only sans the Groom! But, our cousins' tribulations were not yet over. One by one, problems arose: starting with the afore-mentioned failed downstairs apartment call. In the midst of their sons' and Dick's sister's arrivals, their kitchen range failed and had to be replaced; their dishwasher failed and required a hard-to-find part; their car was broken into and although not stolen, had its wiring ripped to shreds; and – tragically- their great granddaughter - a lovely red-headed 11 year old - was diagnosed with a severe case of anorexia. Dick said to me, "And you think Job had troubles?" But, plans for the party continued.

During Susie and Jack's 12 day visit, we saw almost everything in Israel that a visitor should see, except Eilat in the deep south, and Jaffa/ Tel Aviv (due to a luggage arrival time-out). We hit the beaches and great sights in Haifa (the B'Hai Temple, Yad Vashem -- the Holocaust memorial, the Carmelite Nunnery and Monastery, the Sculpture Garden, the Clandestine Museum - where Dick is pictured as one of the founders of the Israeli Navy in 1949 -; the Technion campus (it is the MIT of Israel); rode the Carmelit, Israel's only subway, which connects Carmel with the harbor waterfront area; the Turkish market there; the artist's colony at Ein Hod, and the nearby Druze villages. We made day trips to Acco (Acre) with its wonderful ancient harbor and long winding colorful shouk (casbah); Rosh Ha'Nikrah at the Lebanese border with its beautiful grottos which rival the Blue Grotto at Capri; Caesarea, replete with its Roman amphitheater and harbor and Crusader castles, and also found time to celebrate my 74th at a dinner at Susie's place, with Susan and Dick as guests. We also celebrated Barak's lopsided victory over Bibi Netanyahou, which pleased our local friends no end.

We then began the Christian part of the tour, starting with a day trip to Nazareth to visit the church of the Annunciation, where Mary's pregnancy was announced, and to Mt. Tabor to visit the Church of the Transfiguration, which marks the site where God announced that Jesus was his son. I tend to take these pronouncements at face value.

During the period we were hanging around Haifa waiting for the luggage, we combined our ride downtown on the Carmelit with a quest for Amman bus tickets. (NOTE, this search is covered in detail in a story, 'No Bus to Amman' in my book, "*A Pilgrim Muddles Through*"). We easily found the travel agency where I was told I could get the tickets, but they advised us to walk   "200 meters down Ha'Atzma Ut (Avenue of Independence)" and look for the terminal near the entry to the Port. We did so, but found no bus terminal. I went into a travel agency right across from the port entry and asked where I could buy a ticket for the bus to Amman. They immediately said, "There is no bus to Amman", and I decided to try again after Jack and Susie left. As for a hotel reservation on the departure Saturday night, there was better news. I asked our landlady, Mrs. Shindler (no relation) if she could help and she directed me to her "wonderful" agent (Ordinarily, Dick's travel agent, who had done us proud on our Jordan trip two years ago probably would have been able to solve our problem but, alas, he had recently died in the course of duty in Albania). A few days later, we got a call from our new agent, who proudly told us that while there was no hotel Alia, she did book us into a hotel "near the airport which had the airport shuttle service" that we had insisted on. I was temporarily placated and elated.

We next started off on our grand tour of the Old Country, for which the Brodsky's are justly noted. First, at the start of a blistering heat wave reminiscent of India, we drove to the ancient mystical town of Zefat (Safed), which was absolutely dead, since it was Saturday. After lunch, we visited the war memorial in the Golan Heights, and ended the day by checking into the Scottish Mission in Tiberias, an old stamping ground and our home for two nights in the Sea of Galilee area. The next day, we started off at the kibbutz Ginnosar, with its wonderful museum and ancient recovered boat; then on to the Mount of Beatitudes where Jesus gave the Sermon on the Mount. We drove higher for a lunch stop at the dude ranch, Vered Ha'galil (Rose of Galilee), where Walter Cronkhite used to hang out during the Syrian war, and then back to the Sea to visit Tagpha, where Jesus performed the miracle of the loaves and fishes, and then to Capernaum, where Jesus met and baptized Peter, the Big Fisherman, and had an early synagogue. Before we called it a day, we went looking for the creaky old bridge across the Jordan River which used to separate Israel from Jordan, but missed it. A new bridge had taken its place. We did find the old one, now blocked off to traffic, and wondered how the wretched old one lane structure had withstood the ravages of time so long.

The next morning, we set off with Jerusalem as our ultimate destiny. We had intended to first go to Hammat Gadar, the site of fabulous Roman baths and a huge alligator farm, but our companions wanted to see Yardenit, the serene place where Jesus was purportedly baptized. It is located where the Jordan flows south out of the Sea of Galilee towards its ultimate destiny - the Dead Sea, and it indeed proved to be very peaceful. So, because of the heat and the hour, we decide to forego Hammat Gadar and instead headed to Bet She'an, a marvelous world class (aye, the equal or better than Ephesus in Turkey and Jerash in Jordan) Roman city relic, still being dug out. The temperature on its main street was 110 degrees! From there, we headed smack dab through the West Bank quite uneventfully to Big J, albeit the road skirted Little J (Jericho). By sheer luck - since we had never entered Jerusalem from this direction- we found our hotel at the Hebrew Union College, hard by the walls of the Old City, without any trouble. Strike up one for the pure of heart!

That evening we walked to the famous Ben Jehuda area with all its indoor and outdoor restaurants. We saw the beautiful lighted walls of the old city from our rooms, and so to bed over the din of a wedding party going on at the patio below. Our first venture in Big J was what we recommend for all visitors - take Bus 99, which picks you up near the King David Hotel (which is a stone's throw from the Hebrew Union). This is an inexpensive, English-narrated 2 plus hour tour that takes you to past all the city's highlights, including the impressive Knesset (Parliament) building and grounds and a 10-15 minute 'stop' at the extensive (NO '4') Yad Vashem complex.. It thus allows you to decide which places you want to see in more detail. After lunch, we entered the Old City via the Jaffa Gate, and headed immediately through the Moslem Quarter to the Christian Quarter and the Church of the Holy Sepulchre, which encompasses the grounds on which Christ was crucified and buried. We hired an Arab guide to take us through the church, since here, for once, I cannot claim to be expert in things of the spirit. Susie and Jack were appalled by the crass commercialism in this most holy area, and were shocked to find several "Stations of the Cross" on Via Dolorosa partially covered by vendor's blankets and/or "T" shirts. We met 'the 'Bride' and Susan's newly arrived granddaughter (their Kentucky son's – Dick and Susan's oldest - daughter) for dinner, and found the 'Bride' in good spirits and only a little shell shocked. Before retiring, we had drinks in the lobby of the King David- another "must" for visitors - since it gives you the ambiance of rubbing shoulders with the world's big shots. Over drinks, Patti and I discussed whether we should send Cousin Dick a bill for the new wedding outfits we had both bought, but the Cleland's dissuaded us for the nonce. I had even bought a tie, despite my solemn vow never to wear one again after retirement! (afternote- a promise that I have broken only once or twice).

The next morning, we met the granddaughter and went on a guided tour of the Old City, visiting the Jewish quarter and the Wailing Wall and the ancient Roman Cardo (mall), the Armenian and Moslem quarters, and ending up in the Christian quarter, redoing the Church of the Holy Sepulchre under, this time, our Jewish guide's interpretations. We had lunch near the Jaffa Gate, and Patti and I left for Haifa, leaving the Cleland's to do Masada and Bethlehem on their last two days.

We had a few days left to do gift shopping, socialize with old friends, hit the beach, and once more try to buy bus tickets to Amman before the big Be'ersheva bash. On Friday, May 28, we caravanned with Dick and Susan to the fabulous Be'ersheva Hilton Hotel in time to make Shabbat dinner at their eldest daughter's place in Omer, a suburb. Here, for the first time we met the members of their family that we did not yet know: The middle son, a research doctor of veterinary medicine in Connecticut, and his wife, and the oldest son, then the well-known manager of Three Chimneys horse farm- which features Seattle Slew at stud, among others-, and his lady friend both from Versailles (pronounced "Versals", to the chagrin of Francophiles worldwide), Kentucky. The entire original wedding party-except the aforementioned groom - was present at the dinner/reception -; the 'bride', of course, and all of Dick and Susan's children except their youngest, Portland, Oregon son and his family, were there; plus grandchildren and great-great grandchildren ! It was fun meeting the "new" family members. The 'horsey' son, who had recently obtained a "Kentucky Colonelship" for Dick, lives near an old Navy buddy of mine, and I resolved to visit him and my other friends in the vicinity (my last college roomie, Dick Allen, lives in nearby Cinncy, along with another Cornell buddy). I busily documented the whole group with my camcorder, using this unprecedented opportunity to the utmost. Alas, the Friday night before leaving Israel, I inadvertently and stupidly left the damn thing in the taxicab that took us from Susan and Dick's to our apartment (we had turned in our rental car earlier in the day, and Dick cannot drive on Shabbat). Thus disappeared the entire documentation of the trip, although Susie and Jack shot a lot of film on their camcorder during our tour, which they have subsequently shared with us at their home in Arizona.

We returned to Haifa the next day, not wishing to go on the Masada trek which we had done several times in previous visits. Our last few days in Israel were spent seeing friends, going to the beach, seeing Susan and Dick when they were not showing their oldest son around, and doing more shopping at the Druze villages. On Thursday evening after dinner in central Carmel, I stopped at the taxi dispatch stand to arrange for an early Saturday morning pickup to take us downtown to get the bus to Amman. When I told the dispatcher the plan, he naturally said, "There is no bus to

Amman", and, " No busses run on Saturday". I thanked him for the information and asked him to humor me. We spent our last evening at Dick and Susan's with their Kentucky son and his lady, and then called for the fateful cab, whose driver, I feel sure, now has a nice camcorder to play with.

The bus ride to Amman was pleasant and uneventful, albeit it took over an hour to go through the King Hussein Bridge border crossing, and we were inexplicably charged $15 each to leave Israel! We were let off in downtown Amman and immediately transferred to a cab to go to the Airport hotel, or so we thought. About ten minutes later, the cabbie left us off at the downtown Amman hotel that I recognized we had stayed at two years prior on our way to Petra. I said that there must be some mistake because Mrs. Shindler's travel agent had promised us differently. I went to the desk and asked if they had shuttle bus service to the airport. "Of course", said the pretty clerk, "the taxis run all the time". Really tee'd off, I asked the Clerk if she would release us from our reservation and call the Alia for us. This was done, and we got to the Alia an hour later with not enough Dinars to pay the cab driver. He drove me to the nearby airport and a local ATM. After a pleasant evening and a great meal at the Alia, we were off to Frankfurt the next morning, with three weeks still left on our journey around the world. In Europe, we planned to see friends in Stuttgart, Heidelberg, Aachen and Paris and its environs.

Since we have always used Frankfurt as a jumping off place in Germany, we normally take the E-Bahn subway to the Hof-BahnHof to get the train to Stuttgart / Esslingen to see our dear friend Gretel. And this time, schlepping our bags through the underground maze, we arrived at the main station about an hour before our scheduled departure. We reveled at this - for one of our favorite doings has always been to go to a favorite stand and get a Bratwurst sandwich and a stein of beer, which we absorb at a table amidst the hurrying crowd. As fast as we could, we rushed to the stand, only to find it had been displaced by a McDonald's (at the other former rival stand was a Burger King!). Later, we found that the Americanization of Europe had extended with equal evil to other sectors. The TRAINS no longer run exactly on time!). We settled for a restaurant in the station for the Bratwurst fix- but, not being available as a sandwich - it was just not the same. We boarded the train and soon found that the first stop- as of a week prior- was the Frankfurt airport! You remember the WW II movie about the Remagen Bridge, "A Bridge Too Far"? Well, we had endured in vain a "Schlep Too Far"! Live and learn !

Our week's stay with Gretel in Esslingen, a lovely suburb of Stuttgart, was pleasant as usual. We stayed in a guest house, die Blauer Bock (Blue Ram), a few steps from her condo. The time went quickly, visiting with Gretel's daughter, Betty, and her 93 year old mother, Oma, and her other family members whom we have known for years. We did little sightseeing, for we have covered most of the surrounding state in past visits. One day, however, we did see an outstanding condo community in nearby Plochingen. It was designed by a famous architect/artist named Frederick Hundertwasser, and it is one of five similar communities in Germany and Austria. There must be about 40 adjoining townhouse - type units, each unique, constructed in a triangular plot with the colorful house fronts facing the large garden -like triangular interior courtyard. A high fanciful tower is in the center of one of the three legs. It, like all the other houses, has a multicolored Daliesque finish and is topped out by golden spheres. Each interior - facing house is similarly outlandishly bedecked and every window is of a different size and shape. Balconies have trees growing on them. We decided that it would definitely be fun place to live!

AIN'T RETIREMENT GRAND?

HNDERTWASSER BUILDING

WE VISITED TWO OF THE RENOWNED HUNDERTWASSER COMMUNITIES IN DARMSTADT AND, ABOVE, PLOCHINGEN: BOTH IN THE ENVIRONS OF FRANKFURT. THE MULTI-COLORED BUILDINGS ARE WILDLY PAINTED IN PASTEL SHADES: PINKS, YELLOWS, TANS, BLUES AND REDS

We left by train on Friday to Aachen. On the way, we made a three hour stop in Heidelberg to see and have lunch with our old friends Max and Nieta Klager. He's a Professor (of Pedagogie) at good old H.U. and she's the wife he met while he was getting his Ph.D at the U of Minnesota, and who now asks him, "Max, what's the English word for ----------?" Max is an author of books on art, which he usually translates into English, and is noted for his championing of the art produced by Down's Syndrome artists- vivid, bursting with color, and tres outre! Then, back on the train to Aachen, riding along the very picturesque Rhine River wine and castle country. It was here that we discovered the awful truth of train systems that are now privatized. We missed our connection at Cologne for Aachen - a heretofore unheard of happening! We boarded the next one a half hour later.

Finally arriving, we were greeted by Joachim (Yochen) Damm, the husband of my former German pen pal, now 'adopted' daughter, Gudrun. We first met at the Apfelwein restaurant in Neu Isenburg, near the Frankfurt airport, many years ago. I agreed then to write her in German, while she wrote to me in English- though she was always more competent than I. At our first meeting, we conversed in our then only common language - French. When we met, Gudrun was studying to be a Judge - a position she later reached. Now, since they moved out of her probationary jurisdiction, and now have two children, she can no longer be a Judge without going through the apprenticeship process. So, temporarily at least, she is a lawyer (Avocat), generally working out of her home. Her twin sister, Barbara, also an Avocat, is her partner and lives in Darmstadt,

Yochen is the financial director of Ericsson's German bureau. It is interesting to note that this giant Swedish communications company does all its business in English. What with the Euro then being approximately the same value as the Dollar, and with English rapidly becoming the mutual

European language (except, of course, in France), it was not hard to visualize a United States of Europe before too long, hastened by the new frenzy of conglomeration (viz, Chrysler/ Daimler-Benz).

Our adopted grandchildren, Marina (10) and Peter (7) greeted us like we had never been away! It was over three years. Mein Deutsche still gives them a lot of trouble- but Marina will start taking English this Fall- and we promised to become pen pals! We visited downtown Aachen and Charlemagne's cathedral in the morning, and went to a very busy fair, in the suburbs, in the afternoon. The children had a ball- with rides and junk food! We ate at a fine restaurant- all in all a lovely day with our "kids". Gudrun and Yochen honeymooned with us in Hermosa Beach; visited us a few years ago, and next (Fall, 2001) will bring Die Kinder (we promised them Disneyland and Universal City).

We took the TGV to Paris - 350 miles in 3 hours! - on Sunday morning and were met at Gare Du Nord by Les Ladoires; Gudrun (another one) and Claude. Gudrun was the office manager when I was Aerojet's "Manager of European Operations" in the late 60's. She was then the brains of the gang and kept me out of trouble. She handles French, English, German, and Spanish with equal facility. Claude's English is not good, and it took a few hours for my French to come back with any fluency. I couldn't stop Sprechening die Deutsche! At the station, I cashed in leftover D-Marks for Francs, but for some reason they could not give me money with my VISA card. I was directed to an ATM, and here began a comedy of errors - still inexplicable - which led me to go from one ATM to another - each time being allowed to take out 200-300 Francs (about $35 - $50) each transaction! After we left our bags off at our hotel, we drove about 50 miles in the direction of Lyon towards their retirement compound in Cudot (they had recently given up their pied-a-terre in Paris), stopping along the way at ATMs. Cudot is on the edge of the Burgundy region and is absolutely serene and sylvan in nature. They have a lovely place with two houses and lots of lawn and flower beds and a great miniature Schnauzer, "Jazz". Cudot is a small isolated village- about 10 houses- which abuts large fields and pastures. It is truly idyllic! That evening, Claude cooked great steaks on a fireplace grill while we reminisced about the old days. Next morning, we drove to the rail station in a lovely old town, Sens, where we roamed through the pedestrian mall and had lunch before boarding the train back to Paris.

We took the Metro from the Gare de l'Est to the stop near our hotel. For the second consecutive time, we had chosen to stay in the Hotel of the City Hospital (Hospitel), which is maintained for the benefit of aus-landers coming to visit hospital patients. The glorious thing about it, besides the modest $80/night - with breakfast - cost, is that it is smack dab next door to Notre Dame, right in the middle of the Boul'Mich action. It is well appointed and comes with a night key for use when the hospital lobby entry is closed. The last time we were there, we turned in the key at the hotel desk, as we hurried to make a very early train to the airport. The lobby was closed, and we couldn't find our way out of the hospital ! I was so panicked that I lost all my French, and the people thought I wanted to be admitted. This time, we were forewarned. When we arrived, we found a message from our old navy friend, Tommy, who was then living in Copenhagen. Despite the fact that he had had a stroke several months ago, we had hoped he and his wife Nancy would join us in Paris- but the note canceling the reservation we had made for them said, "No".

That night, we made our first pilgrimage to our favorite restaurant "L'Entrecote", also known as "Les Relais de Venise", just off of the Pt. Maillot stop of Metro #1 line. Our mouth watering was not in vain - the standard and only fare-steak (saignant-rare) with the fabulous mustard sauce, les pommes-fritte, the walnut salad and the red house wine were, as usual, magnificent! After dinner, we walked to our old Monoprix super market to buy some liquid 'supplies'. On the Metro back, we both agreed that it was great to be back in Paris, where we had lived in 1969-70.

The bells of Notre Dame - a stone's throw from our window - woke us up next morning. After petit dejeuner in our room, we Metroed to Le Bourget to see the Paris Air Show; my first appearance there since June, 1969! The big difference the intervening years made was the appearance of booths, companies, and exhibits dealing with things spacial, where before it had been almost exclusively things aeronautical. Nevertheless, I was amazed at the huge variety of aircraft that were exhibited from an equally huge number of countries. The fly-bys- generally on ten minute centers- were both ear busting and daring: Airplanes going straight up; Helicopters doing loops, and, the day before, so we were told, a Russian aircraft barreling in with its two pilots safely ejecting just before hitting the ground. We visited the Aerojet booth and I met the man who was now their European Manager - but his job, selling- was much different than mine was 30 years ago. We also visited the Launchspace booth looking for its president and old friend, Marshall Kaplan (He has sponsored my seminars on remote sensing in the US and abroad). We left him a note to have dinner with us, and returned to our hotel exhausted and standing by for a rest.

We met Marshall that evening at the great Jarrasse sea food restaurant near our former home in Neuilly-sur-Seine, a near suburb at the Pont de Neuilly stop of line #1. Just as we were about to go in, I got a wild hair to try to find another old favorite, the "Le Pied dans L'Eau" (Foot in the Water) on the nearby Ile de La Jatte, an island in the Seine River which starts at the pont and used to contain only playing fields and a few restaurants, accessible by foot only. 'Le Pied' had been a picturesque dive that featured the world's greatest moules. By now, however, the Ile was thoroughly developed with high rises and streets and automobiles. The restaurant, which may have been at the same site, had been rebuilt into a slick very expensive bistro, which did not have moules, at least not this night. We were too tired to walk back to Jarasse and settled for some nondescript appetizers and called it a night.

On Wednesday, we decided to do something new - a 3 mile boat ride down the St. Martin canal which ended at the old Napoleonic arsenal at the Bastille. It was fun ride, past the new Science Museum through the heart of the city. There were three 2-level locks on the way, and the last half mile was through a scary underground grotto. After lunch near the marina at the Port de L'Arsenal, we Metroed back to Neuilly to see our old neighborhood and take a walk in the adjacent Bois de Bologne. Out former house, 5 Ter, Rue de la General Henrion Bertier, was as lovely as ever, although its garage appeared to have been converted, and the building itself appeared to be attached to the building behind it. Perhaps it was now a 3-4 family condo? After the Bois, we went back to the Monoprix to stock up our "water" bottles for tomorrow's trans-Atlantic flight. We agonized as to where we should have our final dinner in Paris- Jarasse or L'Entrecote ?

We wanted to have a go at the wonderful Belon oysters (when we lived there, I would frequently pick up a dozen at the stand outside Jarasse on my way home, and soon became a skilled shucker). We decided that we could always get good sea food in the harbor across the street at home, and opted for L'Entrecote, with a prior stop for oysters at the sea food place near it. It turned out to be an expensive decision- the dozen oysters and glasses of wine cost $50- but what the hay- it was our last night in Paree!

We had a 2 p.m. Icelandic flight to Boston (via a stop in Reykjavik - which, at least from the air and in the vicinity of the airport, didn't even look like a nice place to visit). We had asked the desk clerk at our Paris hotel to arrange for a shuttle to take us to Charles de Gaulle airport. This is a new service which the clerk said was very dependable. To make sure, we asked for a 10:30 pickup, and did not panic until 11:30 - at which time we got a cab (Never thinking of using the rail line whose station was right beneath us- though it would have been hard to get the baggage down the subway steps).

We were met at Logan by son Jeffrey, and his entourage - medium pregnant Lori; their now very voluble and adorable daughter, Emily Ann; and their huge Akita dog, Kobe, and whisked off to Swampscott for our final week. This happy visit at their new home included several trips to the beach (the weather was great!); a few bouts with lobsters and clams; an evening at Fenway Park, watching the Sox beat the Rangers and finally seeing the "Green Monster" wall in person; reunions with old grad school -and-since friends, the Brooks; and a lovely Sunday excursion to a local horse show and a polo match (for me, the first such since my New York days in the late 40's when we used to watch the several extant 10 goal players in Westchester county go at it on the weekends). But, the most fun was playing and talking with Miss Emily - our only girl grandchild- soon, any day now as I finish this epic on August 12, 1999 - to be joined by a sister whose name probably will be Caroline.

We were glad to get home- and were left to remember the highlights with great fondness: The Taj; the cancelled wedding; finding the bus to Amman; the camcorder loss; seeing our relatives and friends in Israel as well as in Germany, France, and Massachusetts. On arrival, we learned of a sad event. "Uncle" (actually first cousin) Harold Brodsky of Philadelphia - the family Patriarch- had died at ~ 83 from a probable after-effect of a back operation - while we were in Paris. I then became the Brodsky family Patriarch.

THE 70TH REUNION

This story, about my ill-fated trip to my 70th high school class reunion, turned out to be nothing like I had anticipated. I had written its beginning paragraphs in the winter of 2011, and it was to be – as it is – the last story in the last chapter of what is probably my last book. My intention was to attend the reunion and finish the story, and thus this book, upon my return. The plan was to depart from Ontario, California to the West Palm Beach airport to visit with friends and relatives before and after the Jan. 27, 2011 event at the Ritz Carlton Hotel in Key Biscayne; returning to Ontario on Feb. 1. My wife, not up to such a strenuous trip, would stay at her cousin's in LaVerne, not far from the Ontario airport, while I was away. A day before we were to make the 40-mile drive inland from our Redondo Beach home, I decided to cancel the trip. I was in some difficulty with heart failure related problems. Three days later, I ended up in the Emergency Room at our local hospital and spent the night. The day after the reunion, back from the hospital, I talked to the Key Biscayne host, a friend from boyhood, and later got more information and pictures from the great event from other attendees. So – even though I was not there – I can tell you something about the event and my classmates who made it there.

Central High School of Philadelphia is the second oldest public high in the country, younger by one year (1834 vs. 1835) than Boston Latin. It, like the Boston school, was authorized by State Assembly to award a Bachelor's Degree to its male-only graduates and accepted only B+ and better students. I was one such and wanted to go there. For years on end, it was rated among the top 20 high schools in the country, and still may be highly rated? It went co-ed about 15 years ago, and that and other factors apparently have affected its ranking status.

It was originally located in downtown Philly – quite a way from my Germantown home – though reachable by a nearly one hour combination of trolley car and subway rides. I chose to wait a year until, in February 1939, the new replacement CHS campus opened much closer (two fairly short trolleys rides) to my home. In the interim year, I went to Germantown High (a long walk) for 9th grade. We, the CHS 177th Class, graduated in January, 1942 – a few weeks after the U.S. entered WW2. I was 16, and because most Universities initiated a three semester-per-year war time schedule, had almost completed my junior year of college when I joined the US Navy in lieu of being drafted.

The first class reunion that I recall was the 30th. It took place at a hotel on City Avenue in Philadelphia across the street from my Mother's condo. Prior to the dinner, she hosted about five of my classmates at cocktails, while we caught up on our news. I had flown in from Iowa, while Obe and Jo came from Washington, DC, Hubert from Boston, Max from California, and Irv was local. The reunion was a great success and we vowed to continue the tradition. The 40th Reunion was held at the same venue as the 30th, and was highlighted by my encounter with classmate Stan Sloan. I reported on this encounter in a story starting on page 16 of my book, "*A Pilgrim Muddles Through*", entitled "The Class Failure". We were all about 57 at the time - a testy age for starting over - so I was devastated when Stan told me "I just lost my job and have no prospects"! What unraveled was that he had just sold his two original NYC franchises for McDonald's shops, plus some other holdings, for several millions (at a time when a Million was a lot of money), and now had nothing to do.

The 60th Reunion – which took place in Philadelphia on September 12, 2001, was a fiasco for me, a Californian, and classmate Ed Swanson, then, as now, a resident of Malmo, Sweden, (see p. 209 in "*Pilgrim—*"). On the fateful morning of September 11, I boarded the 8:30 morning train at Worcester, Mass. to go to Philadelphia, via NYC. Our youngest son, who lives in nearby Boylston, took me to the train station. We were summarily dumped off at New Haven around Noon – where we found out what had happened around 9 o'clock. Ed was similarly stranded in Orlando, where he had been visiting his son on the way to Philadelphia. There was no way I could get to Philadelphia – all rental cars had been grabbed up and all bus service halted. Obviously an Al Qaeda plot to thwart our very lives

M. Dienes

HU, DAVE, AND I AT THE '65TH' IN FLORIDA

AIN'T RETIREMENT GRAND?

DEAR CLASSMATES

Once again the time has come to give notice of our plan for our 69th REUNION, one year ahead of the normal 70th.

As we did in our 65th reunion, we are planning to have a Luncheon at the Hilton Hotel on City Line Ave on October 3rd, 2011 and one at the Ritz Carlton Hotel on Key Biscayne, Florida on January 27, 2011. "Obe" is inviting all classmates and their spouses to lunch and some afternoon activities (a swim or tennis) as his guests for the afternoon.

The Hilton, 4200 city Ave. Philadelphia, Pa 19131 - 215 879 4000

Cash Bar — Ice Tea (No Charge)
Choice of Chicken, Fish - $ 55.00 class members - $ 45.00 others.
Valet Parking
Special Hotel rates will be available for overnight guests.

Each function will commence at 12:00 Noon for Cocktails and Getting Reacquainted, followed by the Luncheon 1 - 4 PM

Let's make this one of the BEST REUNIONS ever. We deserve it !!!

PLEASE COMPLETE THE INFORMATION SHEET AND RETURN IT ASAP INDICATING THE LOCATION YOU PLAN TO ATTEND.

CHS - 177th 69th (Almost 70th) REUNION

My wife and I did make a 65th reunion, one branch of which was held in Boynton Beach, Florida, the winter home of several of my classmates in January, 2007. It was fun seeing them. Milt, an old friend, promised then and there that he would try for a 70th – which he did pull together, albeit a bit early. I vowed to make the 70th, come hell or high water, and was pleased when the notice arrived in early Fall, 2010.

He elected the early date because old friend Obe, wintering on-vacation at the Ritz Carlton Hotel in Key Biscayne, offered to host the luncheon-smoozing event for those of our classmates (actually numbering about half of the ~ 20 survivors) who normally spend the winter in Florida. It seemed smarter to do it earlier, rather than wait a year in view of us survivors all being well into the octogenarian phases of our lives. Milt also planned a second 70th reunion in October, 2011 in Philadelphia. Irv had done the same in getting us organized for the 65th, thus setting a precedent.

Obe and I go way back – to 3rd grade at the C.W. Henry elementary school, which was within walking distance of both our houses in the Germantown suburb of Philadelphia. We also went to dancing class together and shared several friends-for-life. His father, a lawyer, was the Superintendent of Public Schools for the City of Philadelphia. Obe chose to go downtown to the old CHS in the year I went to co-ed Germantown High. After graduation, he went to Dartmouth and did his service in the Army in Europe, and a few years ago wrote his first book about his experiences and attending the postwar Nuremberg trial. Lately, he has written a book telling of his friendship with Supreme Court Justice William Rehnquist (by H.J. Obermayer). He was an attendee in Washington when my technical society dubbed me a 'Fellow'. Sadly, his wife of many years recently has 'disappeared' into Alzheimer's. He nevertheless took her on vacation from his Arlington, Virginia home (He was Editor and Publisher of the *Northern Virginia Sun* - DC's major afternoon newspaper) to Florida and decided to host the Southern-Branch Reunion. With the blessing of Milt, our 177th Class representative, who organized the affair, he sent out, in part, the following letter on Dec. 27, 2010 to me and Hu, Milt, Irv, Leon, Len, Dave, Marv:

M. Dienes

THE 70TH REUNION THAT I MISSED (WITHOUT WIVES WHO ATENDED):

MILT, IRV, MARV, DAVE, OBE, HU, LEON AND LEN : MILT, IRV, DAVE AND LEN ALSO ATTENDED THE OCTOBER REUNION IN PHILADELPHIA

CATCH A ROCKET PLANE

"Dear Classmate:

This is just a brief note to tell you how happy I am that at least eight classmates and six spouses will be attending the celebration of the 69th anniversary of the 177th Class's graduation (almost to the exact date) from Central High School. The festivities will begin on Thursday, January 27, 2011 at the Ritz Carlton hotel on Key Biscayne, part of metropolitan Miami. After lunch, you are welcome to hang around the beach and pool as long as you wish. -----

We share much; a remarkable secondary education, and the privilege of living the best years of our lives in the American Century, among many other things."

In a phone conversation we had after the event, Obe told me that the menu featured Mango soup, lamb chops, and a souffle for desert; all delicious! He commented that most had been married for almost 60 years, except Dave, a Lawyer, who had a 'new' wife. Leon, who had a PhD degree, was a famous endocrinologist with 350 papers to his credit; while Len, a Doctor who specialized in pharmaceutical research, had written around 250 papers (*I, sometimes an Academic, can only claim maybe 150 papers***). Needless to say, he was pleased with the accomplishments of our classmates – all of whom had led exemplary lives.**

After I started feeling better, and living comfortably with my now diagnosed 'congestive heart failure' but still morose about missing out on the great event, I wrote a letter to the attendees:

Feb. 1, 2011

"Dear Classmates –
    Obe, Hu, Milt, Irv, Leon, Len, Dave, Marv:

AIN'T RETIREMENT GRAND?

I was so disappointed that a heart setback caused me to cancel my Reunion trip. I was also going to see old friends and relatives in Palm Beach, Boynton, Boca, N. Miami and Naples. My wife, who has more serious heart troubles, was not up to making the trip.

Just before my scheduled departure, I found myself winded after exertion. I was sure it was a heart artery blockage (I have 4 stents to date), but the Angiogram results (which I had had the previous week) just showed that I was carrying a lot of excess water and my heart wasn't handling it. I spent an over-night at the Hospital while they disposed of the excess water. Now, I'm pretty much back to normal, but with a weaker heart and some new heart medicine to regulate my heartbeat. I plan to resume my normal twice-weekly excursions on my Columbia 36 sailboat on Wednesday with my 'Old Farts' crew (mostly young whippersnappers in the 60s and 70s). BUT- damn-it-all – I did miss the big event and seeing and talking to all of you and the wives that were present and swapping sea stories. I have maintained lifelong relations with Obe (we were in elementary school together), and Hu - who was in the same aerospace engineering profession as I), and, of late, Milt.

I have now received a good run-down on the great event – via pictures and a note from Milt, 2 telephone calls with Obe, letters from Len and Marv and an email from Hu. It sounds like a great time was had- and I am jealous! I doubt if I'll try for the Philly version- since I really have no family left there. But I did get enough info to finish the last story in the last chapter of what very well might be my last (5th) book. The story, which I had started, is called *"The 70th Reunion"*- and it will be different from the one I anticipated, but just as good- hopefully. To make up for my absence, I'll give you a short run-down of my activities, as well a list (enclosed) of my books. Those who are on-line, go to Amazon.com/

books under my name for a short Bio and a detailed look at the books. (*Songs* --- is no longer available).

Until I started writing in the early 1990s, I spent my time, after WW2 service in the Navy, in grad school and – in equal doses- tho sometimes overlapping - in Industry : working on the atom bomb and then the space business, and as a university professor (Iowa State Univ. in the 70s as head of the Aero E Department and in the Astronautical Engineering Dept at USC (1980—'96). I was a pioneer in the space business – in on the ground floor. I worked for Aerojet (I managed its European Office in Paris for a bit) and for TRW (now Northrop Grumman). For a nonce – before and in grad school, I was an unsuccessful musician in Greenwich Village (see, "*The World in a Jug*"). My wife and I (just had 52nd anniv.) now live in King Harbor in Redondo Beach, Calif. Our 4 kids are in Santa Fe, Boylston, Mass and Honolulu. We have 8 grandkids.

My best to all of you and your families. See you at the 75th! "

On February 2, Milt wrote a "Dear Classmates" to the attendees, sending me a copy:

"Phase I of the 177th Class Reunion is now history, with Phase II to be held on October 3, 2011 at the Hilton Philadelphia on City Avenue.

The gathering at the Ritz Carlton, Key Biscayne, couldn't have been better. A beautiful day, a gracious host – Obe – and a superb Luncheon followed by a look-back into the lives and happenings of all of us. The fellowship will be remembered for years to come. It was a GREAT day.

AIN'T RETIREMENT GRAND?

As the unofficially designated Class photographer, I have enclosed -----(see earlier pix)

We all share something "special" as being a part of the Central High School history and tradition that has enriched our lives and rekindled these meaningful friendships.

With fond memories, -------"

Before the trip, I was looking forward to again seeing Hu and his wife, Lori, whom we regularly used to visit on our Florida excursions. He lived near the West Palm Beach airport where I was to land. Both being in the aerospace business, Hu and I had much in common. He, too, had an ScD degree (his from a university in Zurich). We easily would shoot the breeze with 'sea stories'. The plan was for a brief get-together in nearby Boynton Beach where Irv, a retired OBGYN lives. Irv was going to drive me to Boca, where we would pick up Milt and his wife and then head South to the Reunion. When this didn't happen, Hu sent me an email in response to my letter:

"Bob
I was equally disappointed that you could not make our mini-reunion. Well, maybe not so mini anymore, since there are fewer of us left. As you probably heard, there were seven of us there, with six wives. David Cohen, Leon Bradlow, Leonard Dreyfus (sic, Dreifus), Milt Dienes, Irv Arno, "Obe" Obermayer, and myself. Obe hosted a five-star luncheon for us all at the Ritz-Carlton and made introductory comments which made for a very comfortable feeling among us all. Of course, he told us of your difficulty and regrets at your not being there. Everyone, including the wives, made brief remarks, and it was most striking that strong feelings were expressed at the value of the education we received

at Central High. Elliot Lester* was particularly mentioned with high regard by the guys, especially Irv. The value of that education was indicated by the stellar careers of each of the attendees. Milt Dienes had been a sales representative and said that his great ambition was to have sold $100 million of products. He didn't quite make it, falling only $4 million short, but still expressed satisfaction at his achievement. Maxine Arno told of how Irv took her to meet Elliot Lester when they became engaged, to get his approval. David Cohen told us how Len Dreyfus (sic) had saved his life many years ago when he had a heart attack and Len stayed with him all night. He also mentioned trying to help big John Wallace some years after graduation when John got into trouble with the law. I recall hearing Henry Schaefer saying something about that, many years ago, at one of our reunions. That comment reminded me of how there was absolutely no racial issue at Central High and how close we felt to every student, regardless of ethnic identity. I mentioned the time that "colored" (as we said then) John Wallace stayed home on Jewish holidays, because he knew that practically no one would be in class that day.

In my remarks, I said that I was extremely impressed by Obe's letters home in *Soldiering for Freedom*, because of the astute observations by someone only 19 to 20 years old. In *Rehnquist*, I said that we learned almost as much about Obe as about Bill Rehnquist. Reading the book made it clear that Obe and I are far apart in the political spectrum, but that, nevertheless, I still felt a great deal of affection for him. I then told them a little about my aerospace career --------"

---

* Dr. Lester was one of our professors. Most of the faculty had Doctor's degrees.

AIN'T RETIREMENT GRAND?

Later in February, in answer to my earlier letter, I received two letters from other classmate attendees. I will quote salient remarks from both:

Marv wrote from his winter home in Florida, noting that in May, he and his wife would move back to Bala Cynwyd, a Philly suburb, to their condo where they stay for half the year during the summer:

"Hello Bob

Didn't write sooner, but I'm 86 and a half and don't move too quickly. You younger people wouldn't understand. I have attended many of our class reunions, and in each of them we have been chanting "He's Coming! He's Coming! At first I thought it was the ecclesiasticals of the class, but after many years I realize they were just mentioning your name. You were quite the guy of the class. Many thought (*fellow student*) Sam Gaffin was not as good looking as you. But that's another story.

When (*fellow student*) Norm Lazofson and I went into the Army in 1943, he went to the Infantry and then signed up for ASTP at Yale for becoming a Lieutenant. He had ROTC at Drexel. I was drafted into the Air Corps (leaving Temple), and schooled at Sioux Falls Army Radio and Operator classes. This was supposed to lead to gunnery in a B17 until I was grounded with a Pionidal cyst, operated on, and spent the rest of my 3 years in various posts including Paris, France, - Orly Field - in AACS. Norman, unfortunately, was sent to the infantry when ASTP closed. He died in the Battle of the Bulge. **(Ed. Note: *I think about 15 of our classmates met a similar fate during the war. I was awarded an 'American Theater' badge for my valiant service in the US Navy,'44-'46, while classmate Samuel Byron Murphy ended his life-***

time Navy career as an Admiral – to show some extremes in the fortunes of war).

I spent over 50 years after graduating from Temple in processing and manufacturing annealed wire for industry. Betty and I will be married 60 years in April. She graduated from Penn State and became a reading teacher in the Catholic Schools. She is Jewish. We have 2 children: son (internist Doctor) and daughter teaches deaf education in California School system in San Diego. We have grandsons.

Sorry about your bad genes But, you can be sure – in October, I will be chanting "He is Coming!! He is Coming!! Regards"

Physician Len Dreifus, who lives in Heathrow, Florida, wrote:

"Dear Bob
I was pleased to receive your letter detailing the recent events in your life. Happily, your cardiac problem is now under control and you can go sailing once again. I hope your wife's cardiac problem is adequately managed as well.
You were clearly missed at the Florida reunion and Obe served as a most genial host to our group who attended.
We were happy to have this small reunion and are looking forward to the Philadelphia contingent in October that Seline and I hope to attend. We frequently travel to Philadelphia and also visit our children who live in New Jersey.

I am still engaged in active clinical practice, volunteering at the several cardiac clinics. In addition, I am an Adjudicator for Merck Pharmaceuticals and work on a new drug, Anacetrapib, which has three times the potencies of the Statins. I enjoy teaching as a Visiting Professor of Medicine at the University of Central Florida new Medical School in Orlando.

Central High School appears extremely well-kept and managed and is still an outstanding high school experience for the academically gifted students. As you know, it is now a Co-ed institution. -----

Seline and I will be celebrating our 53rd wedding anniversary. We both stay very active and happy other than some minor 'old age' issues. I still enjoy playing my trumpet. Seline plays the piano. We are actively involved with the cultural scene in Orlando.

Keep up your writing and literary interest! We are looking forward to seeing both of you at the next reunion. ----"

It turned out that I did NOT attend the Oct. 3 second 70th reunion, as I was again in the hospital with pneumonia. Right after it occurred, Milt wrote me:

"The 177th 70th Reunion is now History and what an event it was. The turn out was a bit better than I anticipated. With the late addition of Joyce and Dave Cohen, Mort Kaufler, and Fred Lehman, the total came to 12 Classmates and 8 wives plus I arranged for Dr. Sheldon Pavel, President of Central and Harvey Steinberg, President of The Alumni Ass'n. to attend. They definitely added to the festivities. and were appreciative of the invitation. They both spoke briefly but were very interesting talking about the "happenings" at Central. I picked up a large banner of CHS and surprised the guys with Crimson and Gold T- Shirts. Needless to say, photos were taken by guess who. ABC channel 6 showed up at my invitation and did a 30 second segment on the 5:00 o'clock News.

It didn't break up until almost 5:00PM. We had a Ball !!! So very sorry you were not with us. That would have been the "icing" on the cake" which we also had. So there you have it. No formal announcement was made about the next one but I believe that three years is more realistic than five. considering ----------

Take care, stay well and let me hear from you every now and again."

It is with sadness and regret that I finish this story – wondering if I will get to see these fine gentlemen and friends again. I truly expect to make it to our 75th (or 73rd, as the case might be)– but how many of we survivors will still be kicking around?

THE END, FOR NOW

INDEX

Dedication – names of people I worked with at Sandia — iii

Chapter 1   THE MAKING OF AN ENGINEER   1

- PROLOG – GETTING THERE   1
- CAN YOU STEER, MAN?   3
- BORN IN TEXAS   7
- LIKE WEBSTER'S DICTIONARY   17
- TRANSITION – GRAD SCHOOL FROLICS   24

Chapter 2   ATOM BOMB STORIES   27

- THERE'LL BE SOME CHANGES MADE   27
- ALBUQUERQUE – 1950   34
- WHAT! NO FLYING SAUCERS?   37
- SETTLING IN AT SANDIA   38
- BECOMING LANDED GENTRY   48
- A PAEAN TO ALAN POPE   54
- SOMETIMES A GREAT NOTION   60
- TURNED OFF AND GONE   63

Chapter 3   THE SPACE AGE COMETH   71

- GETTING TO CALIFORNIA   72
- IN SPACE, AT LAST!   74
- AN AMERICAN IN PARIS   77
- SPACE (MIS) ADVENTURES IN THE 60s   88
- ARRIVEDERCHI, PAREE!   101

Chapter 4 LA VIE ACADEME   105

- SPACE A'BORNING IN IOWA   105
- A ROCKET RACKET   107
- OVERCOMING FEAR AND TREMBLING   111
- REVEILLE AT REVELLE   114
- LEAVING TENURE BEHIND   117

Chapter 5 SPACE BY THE SEA   119

- LIFE AT TRW   119
- ASTRONAUTICS AT USC   123
- BEATING A DEAD HORSE   130
- A SMASHEROO - WAITING TO HAPPEN   139
- SOLAR SAILING AT THE TECHNION   142

Chapter 6 SNAPSHOTS FROM THE TURN   147
OF THE CENTURY

WE ARE ECLIPSED!   Michael S. Kelly, CEO of KST   148
RUSSIAN MOONY TUNES   Jim Harford,
past AIAA Director   149
A TRW LIGHT & ZAP SHOW   Dr. Tom Romesser of TRW   150
THE AEROSPACE FORCE IS WITH US
General B. Randolph   152
SOJOURNER BEWARE ! Dr. Val Chobotov,
Aerospace Corp.   154
WHAT A GREAT CLAMBAKE IT WAS ! panel discussion   156
QUO VADIS LOGISTICS ? Gerry Facon,
Lockheed Martin Tech Ops   160
WHAT DONE IN TWA 800 ? Prof. Joe
Shepherd of Cal Tech   164

INDEX

THE TRUE APOLLO 13 STORY Owen Brown, AIAA Dist. Lect.	166
COME FLY WITH ME Patrick Carey, *Gulfstream* test pilot	168
THEY'RE BACK !!!! Gordon Ow, CEO, GO Aircraft, LTD.	170
GARDNER GIVES SPACE SKINNY Col. William Gardner, SMC	172
THE AIR FORCE HAS A FULL PLATE! William Maikisch, SMC	173
A HYPER EXPERIENCE Dr. Kevin G. Bowcutt, Boeing	175
A GEM IN THE WEST, JPL's USAF Lt. General (ret.) Gene Tattini	179
THE MISSING LINK by Col. James Painter, SBR Progr. Dir., SMC.	183
TOUGH TIMES IN THE GLOBAL LAUNCH INDUSTRY -J. Schnaars, Boeing Launch Systs.	185
WHO'DA THUNK IT ABOUT "SKUNK"? "Skunk" Baxter, Consult.	188
ANYONE FOR ROCKET TESTING? Col. J. F. Boyle, Comm. AFRL	191
(HEIGH-HO)^2 -IT'S OFF TO MARS WE GO ! Dr. Mark Adler, JPL	193
ARE WE OUT-SOURCING ENGINEERING AND SCIENCE?, Prof. A. Zaragozian, UCLA	196
GPS THEN, NOW, AND THE FUTURE Col. Rick Reaser, SMC	198
THE MISSILE DEFENSE AGENCY HAS A PLAN! Col. Chris Pelc, SMC	201
SBIRS BRIEFING REPORT Col. Scott C. Larrimore, SMC	204
JPL ROLE IN NASA MISSION Gen. Gene Tattini, JPL	206

Chapter 7 FINAL WORKING YEARS — 209

- VISITING PROFESSORING AT THE TECHNION — 209
- THE START OF THE ISRAELI NAVY — 213
- ABET ADVENTURES — 221
- POWER POLITICS AND FUSION — 231
- RETIRING WITH RELUCTANCE AND A LAW SUIT — 237

Chapter 8 AIN'T RETIREMENT GRAND? — 247

- EXPERT WITNESSING AND CONSULTING — 248
- A MEDIA DARLING — 260
- MY LOVE-HATE RELATIONSHIP WITH 'MACHINES' — 277
- AROUND THE WORLD – SPRING '99 — 286
- THE 70th REUNION — 316

PICTURE LISTING

Description	Page
Rocket Plane Tile	i
STEARMAN Biplane	7
Stu Sinclair	10
Ewell & Jimhoit	12
The LINK Trainer	21
Alan Pope	57
The Vaughn's- Hal and Mary Jo	69

INDEX

NERVA nuclear rocket engine	93
Moon Walker, Walking Wheelchair	95
Booster Recovery by Paraglider	100
Aerobee Engine Test Stand	108
Letter from USC President	124
Space Rescue 'lifeboat' Paraglider	134
Interview for TV show on UFOlogy	262
Visiting the Taj Mahal	294
The Hundertwasser Building	309
With classmates at our 65th Reunion	318
Attendees at the 70th Reunion	321

www.ingramcontent.com/pod-product-compliance
Lightning Source LLC
Chambersburg PA
CBHW020726180526
45163CB00001B/122